Overlay Networks

Toward Information Networking

Overlay Networks

Toward Information Networking

Sasu Tarkoma

CRC Press
Taylor & Francis Group
Boca Raton   London   New York

CRC Press is an imprint of the
Taylor & Francis Group, an **informa** business
AN AUERBACH BOOK

Auerbach Publications
Taylor & Francis Group
6000 Broken Sound Parkway NW, Suite 300
Boca Raton, FL 33487-2742

© 2010 by Taylor and Francis Group, LLC
Auerbach Publications is an imprint of Taylor & Francis Group, an Informa business

Printed in the United States of America on acid-free paper
10 9 8 7 6 5 4 3 2 1

International Standard Book Number: 978-1-4398-1371-3 (Hardback)

Library of Congress Cataloging-in-Publication Data

Tarkoma, Sasu.
   Overlay networks : toward information networking / Sasu Tarkoma.
      p. cm.
   Includes bibliographical references and index.
   ISBN 978-1-4398-1371-3 (hardcover : alk. paper)
   1. Information networks. 2. Peer-to-peer architecture (Computer networks) 3. Electronic data processing--Distributed processing. 4. Multimedia communications. I. Title.

TK5105.5.T366 2010
004.6'52--dc22
                                          2009046412

Visit the Taylor & Francis Web site at
http://www.taylorandfrancis.com

and the Auerbach Web site at
http://www.auerbach-publications.com

Contents

Preface

Data and media delivery have become hugely popular on the Internet, with well over 1 billion Internet users. Therefore scalable and flexible information dissemination solutions are needed. Much of the current development pertaining to services and service delivery happens above the basic network layer and the TCP/IP protocol suite because of the need to be able to rapidly develop and deploy them.

In recent years, various kinds of overlay networking technologies have emerged as an active area of research and development. Overlay systems, especially *peer-to-peer* systems, are technologies that can solve problems in massive information distribution and processing tasks. The key aim of many of these technologies is to be able to offer deployable solution for processing and distributing vast amounts of information, typically petabytes and more, while at the same time keeping the scaling costs low.

The aim of this book is to present the state of the art in overlay technologies, examine the key structures and algorithms used in overlay networks, and discuss their applications. Overlay networks have been a very active area of research and development during the last 10 years, and a substantial amount of scientific literature has formed around this topic.

This book has been inspired by the teaching notes and articles of the author in content-based routing. The book is designed not only as a reference for overlay technologies, but also as a textbook for a course in distributed overlay technologies and information networking at the graduate level.

About the Author

Sasu Tarkoma received his M.Sc. and Ph.D. degrees in Computer Science from the University of Helsinki, Department of Computer Science. He is currently professor at Helsinki University of Technology, Department of Computer Science and Engineering. He has been recently appointed as full professor at University of Helsinki, Department of Computer Science. He has managed and participated in national and international research projects at the University of Helsinki, Helsinki University of Technology, and Helsinki Institute for Information Technology (HIIT). He has worked in the IT industry as a consultant and chief system architect, and he is principal member of research staff at Nokia Research Center. He has over 100 publications, and has also contributed to several books on mobile middleware.

Ms. Nelli Tarkoma produced most of the diagrams used in this book.

1

Introduction

1.1 Overview

In recent years, various kinds of overlay networking technologies have emerged as an active area of research and development. Overlay systems, especially *peer-to-peer (P2P)* systems, are technologies that can solve problems in massive information distribution and processing tasks. The key aim of many of these technologies is to be able to offer deployable solution for processing and distributing vast amounts of information, typically petabytes and more, while at the same time keeping the scaling costs low.

Data and media delivery have become hugely popular on the Internet. Currently there are over 1.4 billion Internet users, well over 3 billion mobile phones, and 4 billion mobile subscriptions. By 2000 the Google index reached the 1 billion indexed web resources mark, and by 2008 it reached the trillion mark.

Multimedia content, especially videos, are paving the way for truly versatile network services that both compete with and extend existing broadcast-based medias. As a consequence, new kinds of social collaboration and advertisement mechanisms are being introduced both in the fixed Internet and also in the mobile world. This trend is heightened by the ubiquitous nature of digital cameras. Indeed, this has created a lot of interest in community-based services, in which users create their own content and make it available to others.

These developments have had a profound impact on network requirements and performance. Video delivery has become one of the recent services on the Web with the advent of YouTube [67] and other social media Web sites. Moreover, the network impact is heightened by various P2P services. Estimates of P2P share of network traffic range from 50% to 70%. Cisco's latest traffic forecast for 2009–2013 indicates that annual global IP traffic will reach 667 exabytes in 2013, two-thirds of a zettabyte [79]. An exabyte (EB) is an SI unit of information, and 1 EB equals 10^{18} bytes. Exabyte is followed by the zettabyte ($1\ Z = 10^{21}$) and yottabyte ($1\ Y = 10^{24}$). The traffic is expected to increase some 40% each year. Much of this increase comes from the delivery of video data in various forms. Video delivery on the Internet will see a huge increase, and the volume of video delivery in 2013 is expected to be 700 times the capacity of the US Internet backbone in 2000. The study anticipates that video traffic will account for 91% of all consumer traffic in 2013.

According to the study, P2P traffic will continue to grow but will become a smaller component of Internet traffic in terms of its current share. The current P2P systems in 2009 are transferring 3.3 EB data per month. The recent study indicates that the P2P share of consumer Internet traffic will drop to 20% by 2013, down from the current 50% (at the end of 2008). Even though the P2P share may drop, most video delivery solutions, accounting for much of the traffic increase, will utilize overlay technologies, which makes this area crucial for ensuring efficient and scalable services.

> A P2P network consists of nodes that cooperate in order to provide services to each other. A pure P2P network consists of equal peers that are simultaneously clients and servers. The P2P model differs from the *client-server* model, where clients access services provided by logically centralized servers.

To date, P2P delivery has not been successfully combined with browser-based operation and media sites such as YouTube. Nevertheless, a number of businesses have realized the importance of scalable data delivery. For example, the game company Blizzard uses P2P technology to distribute patches for the *World of Warcraft* game. Given the heavy use of network, P2P protocols such as BitTorrent offer to reduce network load by peer-assisted data delivery. This means that peer users cooperate to transfer large files over the network.

1.2   Overlay Technology

Data structures and algorithms are central for today's data communications. We may consider circuit switching technology as an example of how information processing algorithms are vital for products and how innovation changes markets. Early telephone systems were based on manual circuit switching. Everything was done using human hands. Later systems used electromechanical devices to connect calls, but they required laborious preconfiguration of telephone numbers and had limited scalability. Modern digital circuit switching algorithms evolved from these older semiautomatic systems and optimize the number of connections in a switch. The nonblocking minimal spanning tree algorithm enabled the optimization of these automatic switches. Any algorithm used to connect millions of calls must be proven to be correct and efficient. The latest development changes the fundamentals of telephone switching, because information is forwarded as packets on a hop-by-hop basis and not via preestablished physical circuits. Today, this complex machinery enables end-to-end connectivity irrespective of time and location.

Data structures are at the heart of the Internet. Network-level routers use efficient algorithms for matching data packets to outgoing interfaces based on prefixes. Internet backbone routers have to manage 200,000 routes and more in order to route packets between systems. The matching algorithms include *suffix trees* and *ternary content addressable memories (TCAMs)* [268], which have to balance between matching efficiency and router memory. Therefore, just as with telephone switches, optimization plays a major role in the development of routers and routing systems.

The current generation of networks is being developed on top of TCP/IPs network-layer (layer 3 in the *open systems interconnection (OSI)* stack). These so-called overlay networks come in various shapes and forms. Overlays make many implementation issues easier, because network-level routers do not need to be changed. In many ways, overlay networks represent a fundamental paradigm shift compared to older technologies such as circuit switching and hierarchical routing.

Overlay networks are useful both in control and content plane scenarios. This division of traffic into control and content is typical of current telecommunications solutions such as the *session initiation protocol (SIP)*; however, this division does not exist on the current Internet as such. As control plane elements, overlays can be used to route control messages and connect different entities. As content plane elements, they can participate in data forwarding and dissemination.

> An *overlay network* is a network that is built on top of an existing network. The overlay therefore relies on the so-called *underlay* network for basic networking functions, namely routing and forwarding. Today, most overlay networks are built in the application layer on top of the TCP/IP networking suite. Overlay technologies can be used to overcome some of the limitations of the underlay, at the same time offering new routing and forwarding features without changing the routers. The nodes in an overlay network are connected via logical links that can span many physical links. A link between two overlay nodes may take several hops in the underlying network.

An overlay network therefore consists of a set of distributed nodes, typically client devices or servers, that are deployed on the Internet. The nodes are expected to meet the following requirements:

1. Support the execution of one or more distributed applications by providing infrastructure for them.
2. Participate in and support high-level routing and forwarding tasks. The overlay is expected to provide data-forwarding capabilities that are different from those that are part of the basic Internet.
3. Deploy across the Internet in such a way that third parties can participate in the organization and operation of the overlay network.

Figure 1.1 presents a layered view to overlay networks. The view starts from the underlay, the network that offers the basic primitives of sending and receiving messages (packets). The two obvious choices today are UDP and TCP as the transport layer protocols. TCP is favored due to its connection-oriented nature, congestion control, and reliability.

After the underlay layer, we have the custom routing, forwarding, rendezvous, and discovery functions of the overlay architecture. Routing pertains to the process of building and maintaining routing tables. Forwarding is the process of sending messages toward their destination, and rendezvous is a function that is used to resolve issues regarding some identifier or node—for example, by offering indirection support in the case of mobility. Discovery is an integral part of this layer and is needed to populate the routing table by discovering both physically and logically nearby neighbors.

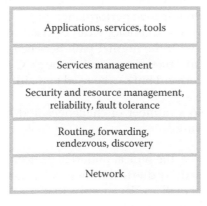

FIGURE 1.1
Layered view to overlay networks.

The next layer introduces additional functions, such as security and resource management, reliability support, and fault tolerance. These are typically built on top of the basic overlay functions mentioned above. Security pertains to the way node identities are assigned and controlled, and messages and packets are secured. Security encompasses multiple protocol layers and is responsible for ensuring that peers can maintain sufficient level of trust toward the system. Resource management is about taking content demand and supply into account and ensuring that certain performance and reliability requirements are met. For example, relevant issues are data placement and replication rate. Data replication is also a basic mechanism for ensuring fault-tolerance. If one node fails, another can take its place and, given that the data was replicated, there is no loss of information.

Above this layer, we have the services management for both monitoring and controlling service lifecycles. When a service is deployed on top of an overlay, there need to be functions for administering it and controlling various issues such as administrative boundaries, and data replication and access control policies.

Finally, in the topmost layer we have the actual applications and services that are executed on top of the layered overlay architecture. The applications rely on the overlay architecture for scalable and resilient data discovery and exchange.

An overlay network offers a number of advantages over both centralized solutions and solutions that introduce changes in routers. These include the following three key advantages:

Incremental deployment: Overlay networks do not require changes to the existing routers. This means that an overlay network can be grown node by node, and with more nodes it is possible to both monitor and control routing paths across the Internet from one overlay node to another. An overlay network can be built based on standard network protocols and existing APIs—for example, the Sockets API of the TCP/IP protocol stack.

Adaptable: The overlay algorithm can utilize a number of metrics when making routing and forwarding decisions. Thus the overlay can take application-specific concerns into account that are not currently offered by the Internet infrastructure. Key metrics include latency, bandwidth, and security.

Robust: An overlay network is robust to node and network failures due to its adaptable nature. With a sufficient number of nodes in the overlay, the network may be able to offer multiple independent (router-disjoint) paths to the same destination. At best, overlay networks are able to route around faults.

The designers of an early overlay system called *resilient overlay network (RON)* [361] used the idea of alternative paths to improve performance and to route around network faults. Figure 1.2 illustrates how overlay technology can be used to route around faults. In this example, there is a problem with the normal path between A and B across the Internet. Now, the overlay can use a so-called *detour path* through C to send traffic to B. This will result in some networking overhead but can be used to maintain communications between A and B.

Overlay networks face also a number of challenges and limitations. The three central challenges include the following:

- The real world: In practice, the typical underlay protocol, IP, does not provide universal end-to-end connectivity due to the ubiquitous nature of firewalls and *network address translation (NAT)* devices. This means that special solutions are needed to overcome reachability issues. In addition, many overlay networks are oblivious to the current organizational and management structures that exist in applications

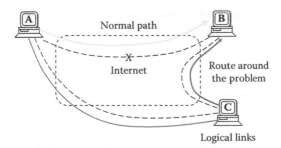

FIGURE 1.2
Improving resiliency using overlay techniques.

and also in network designs. For example, most of the overlay solutions presented in this book do not take Internet topology into account from the viewpoint of the *autonomous systems (ASs)* and inter-AS traffic.

- Management and administration: Practical deployment requires that the overlay network have a management interface. This is relatively easy to realize for a single administrative domain; however, when there are many parties involved, the management of the overlay becomes nontrivial. Indeed, at the moment most overlays involve a single administrative domain.

  The administrator of an overlay network is typically removed from the actual physical devices that participate in the overlay. This requires advanced techniques for detecting failed nodes or nodes that exhibit suspect behaviors.

- Overhead: An overlay network typically consists of a heterogeneous body of devices across the Internet. It is clear that the overlay network cannot be as efficient as the dedicated routers in processing packets and messages. Moreover, the overlay network may not have adequate information about the Internet topology to properly optimize routing processes.

Figure 1.3 presents a taxonomy of overlay systems. Overlays can be router-based, or they can be completely implemented on top of the underlay, typically TCP/IP. Router-based overlays typically employ IP Multicast [107, 130] and IP Anycast [106] features; however, given the fact that deployment of the next version of the IP protocol, IPv6 [106], has not progressed according to most optimistic expectations, these extensions are not

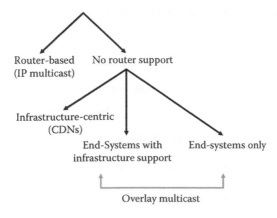

FIGURE 1.3
Taxonomy of overlay networks.

globally supported on the Internet. If the routers only provide basic unicast end-to-end communication, information networking functions need to be provided by the overlay.

> *Content delivery networks (CDNs)* are examples of overlay networks that cache and store content and allow efficient and less costly ways to distribute data on a massive scale. CDNs typically do not require changes to end-systems, and they are not P2P solutions from the viewpoint of the end clients.

The two remaining categories illustrated in Figure 1.3 are end-systems with and without infrastructure support, respectively. The former combines fixed infrastructure with software running in the end-systems in order to realize efficient data distribution. The latter category does not involve fixed infrastructure, but rather establishes the overlay network in a decentralized manner.

Overlay networks allow the introduction of more complex networking functionality on top of the basic IP routing functionality. For example, *filter-based routing*, *onion routing*, *distributed hash tables (DHTs)*, and *trigger-based forwarding* are examples of new kinds of communication paradigms. DHTs are a class of decentralized distributed algorithms that offer a lookup service. DHTs store (key, value) pairs, and they support the lookup of the value associated with a given key. The keys and values are distributed in the system, and the DHT system must ensure that the nodes have sufficient information of the global state to be able to forward and process lookup requests properly.

The DHT algorithm is responsible for distributing the keys and values in such a way that efficient lookup of the value corresponding to a key becomes possible. Since peer nodes may come and go, this requires that the algorithm be able to cope with changes in the distributed system. In addition, the locality of data plays an important part in all overlays, since they are executed on top of an existing network, typically the Internet. The overlay should take the network locations of the peers into account when deciding where data is stored, and where messages are sent, in order to minimize networking overhead.

Figure 1.4 illustrates the key DHT API functions that allow peers to insert, look up, and remove values associated with a key. Typically, the key is a hash value, so-called *flat label*, which realizes essentially a flat namespace that can be used by the DHT algorithm to optimize processing.

> DHTs are a class of decentralized distributed systems. They provide a logically centralized lookup service similar to hash tables. A DHT stores (key, value) pairs and allows a client to retrieve a value associated with a given key. The DHT is typically realized as a structured P2P network in which peers cooperate to provide the service across the Internet.

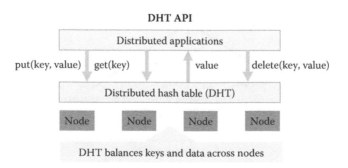

FIGURE 1.4
DHT API.

There are two main classes of P2P networks, *structured* and *unstructured*. In the former type, the overlay network topology is tightly controlled by the P2P system and content is distributed in such a way that queries can be made efficiently. Many structured P2P systems utilize DHT algorithms in order to map object identifiers to distributed nodes. Unstructured P2P networks do not have such tightly controlled structure, but rather they utilize flooding and similar opportunistic techniques, such as *random walks* and *expanding-ring time-to-live (TTL)* search, for finding peers that host interesting data. Each peer receiving a query can then evaluate the query locally using its own content. This allows unstructured P2P systems to support more complex queries than are typically supported by structured DHT-based systems.

Unstructured P2P algorithms are called *first generation* and the structured algorithms are called *second generation*. They can also be combined to create *hybrid* systems. The key-based structured algorithms have a desirable property: namely, that they can find data locations within a bounded number of overlay hops [162]. The unstructured broadcasting-based algorithms, although resilient to network problems, may have large routing costs due to flooding, or may be unable to find available content [274].

Another approach to P2P systems is to divide them into two classes, *pure* and *hybrid* P2P systems. In the former, each peer is simultanously a client and a server, and the operation is decentralized. In the latter class, a centralized component is used to support the P2P network.

Figure 1.5 illustrates the inherent trade-off between completeness and expressiveness of an overlay system. By completeness we mean the ability of the system to guarantee the location and retrieval of a piece of data. Expressiveness pertains to the system's ability to reason about the data—for example, how complex queries can be used to locate data elements. DHTs and other structured overlays typically guarantee completeness, whereas unstructured systems, such as Gnutella and Freenet, do not provide such guarantees. As an inherent limitation, structured systems support less complex queries, typically the lookup of keys. Unstructured systems, on the other hand, can support complex query processing. In this book, we cover both structured and unstructured systems and highlight their key properties.

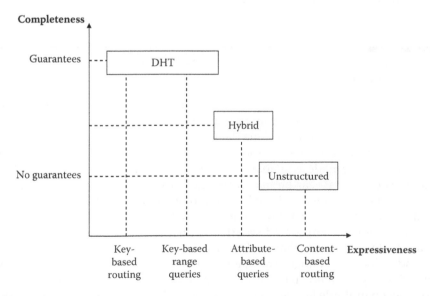

FIGURE 1.5
Balancing completeness and expressiveness in overlays.

1.3   Applications

Many overlay networks have been proposed both in the research community and by Internet and Web companies. Overlay networks can be categorized into the following classes [80]:

- P2P file sharing: For sharing media and data. For example, Napster, Gnutella, KaZaA.

- CDN: Content caching to reduce delay and cost. For example, Akamai and LimeLight.

- Routing and forwarding: Reduce routing delays and cost, resiliency, flexibility. For example, resilient overlay network (RON), Internet indirection infrastructure (i3).

- Security: To enhance end-user security and offer privacy protection. For example, virtual private networks (VPNs), onion routing, anonymous content storage, censorship resistant overlays.

- Experimental: Offer testing ground for experimenting with new technologies. For example, PlanetLab.

- Other: Offer enhanced communications. For example, e-mail, VoIP, multicast, publish/subscribe, delay tolerant operation, etc.

Currently a significant amount of content is being served using decentralized P2P overlays. Most of the deployed algorithms are based on unstructured overlays. The unstructured P2P protocol BitTorrent has become a popular content distribution protocol over the recent years.

P2P technologies are not commonly used with CDNs; however, they are increasingly used by end clients. P2P offers end client–assisted data distribution, in which clients acting as peers upload data. This contrasts with the traditional client-server CDN model, in which clients do not upload data. The main strength of P2P is in the delivery of massively popular data items; however, items that fall into the long tail may not be cost-efficient to distribute using P2P. This can be alleviated by storing data items on client machines using caching, but this requirement is not favored by many users.

1.4   Properties of Data

In this section, we briefly discuss the properties of data [117, 120, 228]. Data can be characterized in many ways. We consider an example taxonomy in Figure 1.6 that divides data into two parts: stored data and real-time data.

Stored data consists of bits that are stored on a system on a more permanent basis in such a way that the data can be made available later. This data can take two forms: it can be mutable or immutable. Mutable data can be shared and modified by various entities either locally or in the distributed environment. Mutable data can be made incrementally available, and it can be created and managed by multiple entities. On the other hand, mutable data is not easy to cache and it requires complicated security solutions, especially in distributed environments. Immutable data means that the full data—for example, a picture or a video file—is available, and it does not change. This data can therefore be cached and verified easily.

Real-time data is generated on the fly and transmitted over the network. The data is packetized, possibly on multiple layers, and it is transferred hop-by-hop on a store-and-forward basis. This means that, although individual packets of the data are stored in intermediate

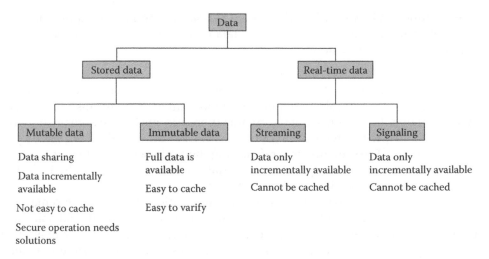

FIGURE 1.6
Taxonomy of data.

buffers, the whole data is not stored as such. In addition, with real-time data, the time when the data is inserted into the network plays a crucial part.

Streaming data is only incrementally available, and only the latest packets of this stream are important. This means that this kind of data cannot be cached. Another form of real-time data is signaling. In this case, data also becomes incrementally available and cannot be cached; however, the data packets are typically very different from streaming.

References play an important part in distributed systems. A reference encapsulates a relationship between itself and a referent defined relative to the state of some physical system. As examples we may consider memory addresses that point to some specific locations of physical memory and *universal resource locators (URLs)* that point to Web resources located on specific servers, available using a specific protocol such as the *hypertext transfer protocol (HTTP)*. If the physical system changes—for example, memory is swapped or a server is relocated—the referent changes as well. These so-called *physical references* may become invalid when the environment changes.

In order to cope with changes in the environment, the common practice is to introduce a level of *indirection* into the reference system. For example, the *domain name system (DNS)* binds host names to IP addresses, which allows administrators to change IP addresses without changing host names. The hierarchical and replicated structure of DNS scales well for its intended purposes, and it is at the core of the Internet.

A data element can be either mutable or immutable. In the former case it can change, and in the latter case it cannot change. It is obvious that a mutable data element can be represented by a sequence (or a graph) of immutable data elements. Given that a piece of data does not change, it can be uniquely and succinctly summarized using a hash function. We note that hashes only provide probabilistic uniqueness; however, a long enough hash bitstring results in a vanishingly small probability of collision.

A hash function is a function from a sequence of bytes to a fixed size sequence of bits, a bitstring. Hash functions can be characterized based on how easy it is to find a collision [227]:

- A hash function is *strongly collision resistant* if it is not computationally feasible to find two different input data items which have the same hash.
- A hash function is *weakly collision resistant* if, for a given data item, it is computationally not feasible to find another data item that has the same hash.

- A hash function is *probabilistically collision resistant* if, for a given input data item, the probability that a randomly chosen data item will have the same hash as the input data item is extremely small.

Semantic-free references have been proposed to achieve persistence and freedom from contention in a naming system [20, 339]. The idea is to use a reference namespace devoid of explicit semantics—for example, based on hashed identifiers. This means that a reference should not contain information about the organization, administrative domain, or network provider. Flat semantic-free references contrast with DNS-based URLs because they have no explicit structure. The semantic-free referencing method uses DHTs to map each object reference to a machine that contains object metadata. The metadata typically includes the object's current network location and other information.

Until recently, there have been no good candidate solutions for resolving semantic-free names in a scalable fashion in the distributed environment. The traditional solution has been to use a partitioned set of context-specific name resolvers. The emerging overlay DHT technology can be used to efficiently store and look up semantic-free references. Indeed, the so-called self-certified flat labels have gained widespread adoption in recent overlay systems.

Self-certifying data is data whose integrity can be verified by the client accessing it [227]. A node inserting a file in the network or sending a packet calculates a cryptographic hash of the content using a known hash function. This hashing produces a file key that is included in the data. The node may also sign the hash value with its private key and include its public key with the data. This additional process allows other nodes to authenticate the original source of the data. When a node retrieves the data using the hash of the data as the key, it calculates the same hash function to verify that the data integrity has not been compromised.

A large part of the research and development on P2P systems has focused on *data-centric* operation, which emphasizes the properties of the data instead of the location of the data. Ideally, the clients of the distributed system are not interested in where a particular data item is obtained as long as the data is correct. The notion of data-centricity allows the implementation of various dynamic data discovery, routing, and forwarding mechanisms [274].

In *content-based routing* systems, hosts subscribe to content by specifying filters on messages. In content-based routing, the content of messages defines their ultimate destination in the distributed system. Information subscribers use an interest registration facility provided by the network to set up and tear down data delivery paths. Data-centric and content-based communications are currently being investigated as possible candidates for Internet-wide communications.

1.5 Structure of the Book

After the introduction chapter that motivates overlay technology and outlines several application scenarios, we start with an overview of networking technology in Chapter 2. This chapter briefly examines the TCP/IP protocol suite and the basics of networking, such as naming, addressing, routing, and multicast. The chapter forms the basis for the following chapters, because typically TCP/IP is the underlay of the overlay networks and thus understanding its features and properties is vital to the development of efficient overlay solutions.

We discuss properties of networks in Chapter 3, including the growth of the Internet, trends in networking, and how data can be modeled. Many of the overlay algorithms are based on the observation that networks exhibit power law degree distributions. This can then be used to create better routing algorithms.

In Chapter 4 we examine a number of unstructured P2P overlay networks. Many of these solutions can be seen to be part of the first generation of P2P and overlay networks; however, they can be also combined with structured approaches to form hybrid solutions. We cover protocols such as Gnutella, BitTorrent, and Freenet and present a comparison of them. This chapter places special emphasis on BitTorrent, because it has become the most frequently used P2P protocol.

Chapter 5 presents the foundations of structured overlays. We consider various geometries and their properties that have been used to create DHTs. The chapter also presents consistent hashing, which is the basis for the scalability of many DHTs. After surveying the foundations and basic cluster-based solutions, we then examine a number of structured algorithms in Chapter 6. Structured overlay technologies place more assumptions on the way nodes are organized in the distributed environment. We analyze algorithms such as the Plaxton's algorithm, Chord, Pastry, Tapestry, Kademlia, CAN, Viceroy, Skip Graphs, and others. The algorithms are based on differing structures, such as hypercubes, rings, tori, butterflies, and skip graphs. The chapter considers also some advanced issues, such as adding hierarchy to overlays.

Many P2P protocols and overlay networks utilize probabilistic techniques to reduce processing and networking costs. Chapter 7 presents a number of frequently used and useful probabilistic techniques. Bloom filters and their variants are of prime importance, and they are heavily used in various network solutions. The chapter also examines epidemic algorithms and gossiping, which are also the foundation of a number of overlay solutions.

As observed in this chapter, data-centric and content-centric operation offer new possibilities regarding data caching, replication, and location. Recently, content-based routing has become an active research area. In Chapter 8 we consider content-centric routing and examine a number of protocols and algorithms. Special emphasis is placed on distributed publish/subscribe, in which content is targeted to active subscribers.

Given the scalable and flexible distribution solutions enabled by P2P and overlay technologies, we are faced with the question of security risks. The authenticity of data and content needs to be ensured. Required levels of anonymity, availability, and access control also must be taken into account. Chapter 9 examines the security challenges of P2P and overlay technologies, and then outlines a number of solutions to mitigate the examined risks. Issues pertaining to identity, trust, reputation, and incentives need to be analyzed.

Chapter 10 considers applications of overlay technology. Amazon's Dynamo is considered as an example of an overlay system used in production environment that combines a number of advanced distributed computing techniques. We also consider *video-on-demand (VoD)* in this chapter. Much of the expected IP traffic increase in the coming years will come from the delivery of video data in various forms. Video delivery on the Internet will see a huge increase, and the volume of video delivery in 2013 is expected to be 700 times the capacity of the US Internet backbone in 2000. The remainder of the chapter examines P2P SIP for telecommunications signaling, and content distribution technologies.

Finally, we conclude in Chapter 11 and summarize the current state of the art in overlay technology and the future trends. The chapter outlines the main usage cases for P2P and overlay technologies for applications and services.

2

Network Technologies

This chapter examines the TCP/IP protocol suite and the basics of networking, such as naming, addressing, routing, and multicast. The chapter forms the basis for the following chapters, because typically TCP/IP is the underlay of the overlay networks and thus understanding its features and properties is vital for the development of efficient overlay solutions. The chapter places emphasis on interdomain routing, because it is key for scalable and policy-compliant global networking. Overlay solutions should ensure that the underlay is used in an efficient and policy-compliant manner [203].

2.1   Networking

TCP/IP forms the basis of the current Internet, and it is generally described as having four abstraction layers—namely, the link layer, network layer, transport layer, and application layer. This layered view is often compared with the seven-layer *OSI reference model*. Design principles, outlined in RFC 1122, have had a major influence on the development of the current Internet [106]. The two key design principles for the Internet were [81] the *end-to-end principle* and the *robustness principle*.

The end-to-end principle places the maintenance of state and overall intelligence at the edges and assumes the core Internet retains no state [282]. Today's real-world needs for firewalls, *network address translation (NAT)*, and web content caches have essentially made this principle impossible to follow in practice.

The robustness principle can be summarized as follows: *be conservative in what you do, be liberal in what you accept from others*. The principle suggests that Internet software developers carefully write software that adheres closely to extant RFCs but accept and parse input from clients that might not be consistent with those RFCs. As stated in RFC 1122, adaptability to change must be designed into all levels of Internet host software.

The network layer in the TCP/IP model is responsible for realizing internetworking and uses the IP protocol to deliver data from upper layers between end hosts. The protocol suite separates host names from topological addresses by using name resolution. The *domain name system (DNS)* is responsible for resolving hierarchical host names to topological IP addresses [231]. This effectively separates naming from addressing, and even though the naming system, namely DNS, fails, the underlying routing can still function independently. DNS also allows the definition of organizational boundaries that are independent of the network topology.

A *routing algorithm* is responsible for building and maintaining routing tables. A *forwarding algorithm* is responsible for determining the next hop given a destination address. Packet routing involves use of routing and forwarding algorithms and protocols for deciding where an incoming packet should be sent. The two main classes are intradomain and interdomain protocols. Intradomain protocols are applied in an autonomous system (AS)—for example, a *metropolitan area network (MAN)* or regional network—and interdomain protocols are used to connect the different AS together to form a global network topology. The typical examples of the protocols are *open shortest path first (OSPF)* for intradomain operation and *border gateway protocol (BGP)* for interdomain operation.

The communications models offered by the Internet can be categorized into the following cases. In *unicasting*, a packet traverses a sequence of links from a source to a destination. The majority of traffic on the Internet is unicast. In *multicasting*, a packet selectively traverses multiple chains of links from typically one source to multiple destinations. In *broadcasting*, a packet is sent on multiple links to every device on the network. In practice, broadcast is applied only within a specific broadcast domain. In *anycasting*, a suitable chain of links is selected from a number of possible candidates. Packets are sent to the nearest or best destination.

Of the above communication models, the currently dominant IP version 4 protocol supports only unicasting on a global scale. The next version of IP, version 6, offers these other communication models as well; however, the IPv6 deployment has not progressed according to some optimistic expectations, and it remains to be seen when the new protocol is globally deployed.

> The Internet is based on hierarchical routing, in which *autonomous areas (AS)* are connected by peering and transit links. Each AS can run its own local routing algorithm, and BGP is used for interdomain connectivity.

Figure 2.1 illustrates the interoperable nature of the IP protocol. The network layer provides global addressing and end-to-end reachability, and thus abstracts the applications from the details of routing and forwarding. The IP protocol supports a number of underlying links and physical layer protocols, which makes it the waist of the protocol stack. Higher-level features diverge from the IP and support different operating environments. The network layer therefore minimizes the number of service interfaces and maximizes interoperability.

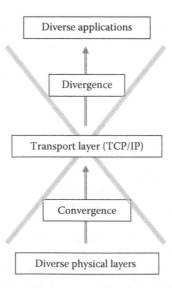

FIGURE 2.1
Hourglass model in networking.

Middleware provides additional services on top of the networking stack and below the applications. Most overlay and P2P technologies can be thought to be part of middleware. As middleware, they utilize the APIs and features of the underlying protocol stack and network and offer their own APIs for application developers. The motivation for this layer is that it can abstract many details pertaining to the underlying layers and thus make it easier to develop and run distributed software.

TCP/IP applications use either a host name or an IP address. The former requires a DNS lookup to resolve the IP address, whereas the latter is directly routable. Recently there have been a number of proposals for adding further indirection into the protocol architecture by means of *locator-identity split*. In general, the split would allow various identifiers—for example, cryptographic identifiers [14, 188, 243]—to be mapped to IP addresses. The motivation for locator-identity split is increased flexibility and de-emphasizing the central role of IP addresses as end-point identifiers.

2.2 Firewalls and NATs

The present-day Internet has seen ubiquitous deployment of firewalls and *network address translators (NATs)*. Both are used to control data communications between subnetworks. Firewalls are hardware or software components that block certain incoming connections. Their main motivation is to increase security by preventing unauthorized connections to a device. NAT devices, on the other hand, perform conversion between different address spaces, typically private and public networks (Fig. 2.2). The motivation for NATs is that they offer a certain level of security and allow the use of private IP address spaces, thus alleviating IP address exhaustion concerns and some network management concerns as well.

> A NAT involves the translation of an IP address used within one network to a different IP address known within another network. Typically, a NAT maps its local private network addresses to one or more global outside IP addresses and then performs reverse mapping of the global IP addresses on incoming packets back into private IP addresses.

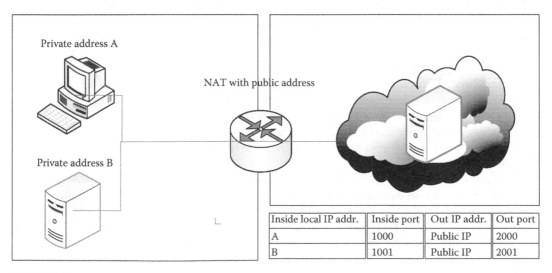

Inside local IP addr.	Inside port	Out IP addr.	Out port
A	1000	Public IP	2000
B	1001	Public IP	2001

FIGURE 2.2
Example of network address translation.

There are a variety of NAT devices and a variety of network topologies utilizing NAT devices in deployments. NAT devices support private IP addressing domains that are not globally reachable. Typically, client-initiated connections create soft state in the NAT devices so that responses to requests can be sent to the hosts in the private domain.

There are four general types of NAT devices, based on how they perform the address mapping:

- Full cone NAT maps an internal address to an external address in one-to-one fashion, and it is easy to traverse.
- Restricted cone NAT maps internal address (and port) to an external address. Once the internal client has sent a packet to an external host, the external host can send packets back from any port.
- Port-restricted cone NAT is similar to the restricted cone NAT, but the external host can only send from the port to which it received packets from the internal client.
- In symmetric NAT, only an external host that receives packets from the internal host can send packets back.

The asymmetric addressing and connectivity domains established by NAT devices have created unique problems for P2P systems, which realize both client and server functionality at end nodes. NATs may prevent P2P nodes from receiving inbound requests. Although P2P systems build on the end-to-end communications capability of the Internet, in practice the assumption that a peer can receive inbound traffic is often not valid.

A number of techniques have been devised for applications to detect the NATs on the communication path and then configure the communications in such a way that the connection can be established. The communication options depend on the type of NATs. The worst case happens when there are symmetric NATs present, which map each outgoing connection to a new IP address and port number. This case is solved by using a special rendezvous server that relays all packets between the communicating endpoints [302].

IETF has developed a number of NAT traversal solutions that include connection establishment (STUN), relaying (TURN), and combined solutions for SIP (ICE). The solutions are surveyed and discussed in RFC 5128 [302]. *Relaying* is the most reliable method of realizing NAT traversal; however, it is also the least efficient, because the relay server's processing power and network capacity is used to relay packets. Another technique is *connection reversal* for direct communication that works if only one of the two peers is behind a NAT device. *UDP and TCP hole punching* can be used to punch holes through NAT devices and establish direct connectivity between peers even when both hosts are behind NATs. Recent analysis results indicate that UDP hole punching works widely on more than 80% of the NAT devices. TCP hole punching is not as frequently supported, with approximately 60% support.

P2P applications may use multiple rendezvous servers for registration, discovery, and relay functions. As an example, Skype uses a central public server for login and a number of different public servers to realize end-to-end relay functionality. Recent studies based on thousands of BitTorrent swarms indicate that roughly half of the peers can be behind firewalls [232]. We return to the Skype and BitTorrent protocols in more detail in Chapter 4.

2.3 Naming

Names and namespaces are fundamental components of network architectures. In the current Internet, the DNS is responsible for managing the hierarchical domain namespace. The DNS protocol was specified in the early 1980s by the IETF. Much of the flexibility of the current Internet stems from the scalability of both network-level hierarchical routing and the higher level naming service. DNS has facilitated the deployment of the World Wide Web and e-mail.

DNS is a managed distributed overlay that uses a static distribution tree and a hierarchically organized namespace. The DNS system is a distributed database system implemented using the client-server model, in which the nameservers are responsible for the sharing, replicating, and partitioning the domain namespace, and answering client requests. DNS achieves scalability and resilience by relying extensively on caching and replication. As a consequence, updates to DNS records typically require some time to become globally available. Another limitation of DNS is that it does not have built-in security, which makes it prone to a number of vulnerabilities.

The client-side uses a DNS resolver to look up information from DNS. DNS uses UDP for typical requests and TCP for larger transfers. The DNS system supports two different query modes, namely nonrecursive queries and recursive queries. A nonrecursive query places the control at the requesting client, and typically a single DNS provides only a partial answer to the query. The client can then expand the partial answer by using other nameservers that are identified in the partial answer. A recursive query, on the other hand, places the control of the resolution process at the nameserver, which will then contact other nameservers to answer the query. This latter mode is not a mandatory feature.

The namespace consists of domain names that are organized in a tree structure. Each domain name in this tree has zero or more resource records that contain information about the name. Each domain name is part of a *DNS zone* and has one or more authoritative DNS servers. The root level of the hierarchy is served by the root nameservers, which are used to look up a *top-level domain name (TLD)*.

A DNS zone consists of a set of nodes served by an authoritative nameserver. Administrative responsibility of a zone can be divided to multiple nameservers. Moreover, a single nameserver can be responsible for multiple zones. Authority can be delegated for an arbitrary part of a zone, typically in the form of subdomains. In the case of delegation, the new nameserver will become the authoritative nameserver for the delegated namespace.

The Internet Corporation for Assigned Names and Numbers (ICANN) oversees the registrar companies that maintain top-level domains. The domain names have a hierarchical structure, and new hierarchy levels can be assigned under the top-level domains. The DNS domain hierarchy is independent of network topology and network administrative domains. This means that multiple names can be mapped to the same network and same physical server. A name can also map to different IP addresses based on some policy, which is useful in realizing *load balancing*. The separation of naming and addressing thus provides flexibility by allowing more fine-grained policies to be implemented.

The DNS service has been designed to accept queries pertaining to host names and IP addresses. A DNS client can perform a lookup to translate a hostname to an IP address, translate an IP address to a hostname, and obtain published information about a host (typically MX record for e-mail SMTP server details).

Figure 2.3 illustrates how DNS is used. When a client needs to obtain information about a hostname, it sends a query to its local DNS server. The local DNS server consults its own

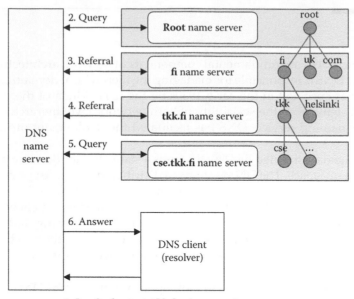

1. Resolve **host.cse.tkk.fi** using recursive query

FIGURE 2.3
Overview of the domain name system.

cache if it already has the answer to the query. If the cache does not contain the answer, the local DNS server forwards the query to other DNS servers. Once the DNS server receives an answer, it can cache it before sending it to the client.

We can take the lookup for cse.tkk.fi as an example. The local DNS server first queries one of the public root nameservers to find the machines that are nameservers for the .fi domain. Then the local DNS server queries the .fi domain nameservers to determine the nameservers responsible for the tkk.fi domain. Finally, it queries the tkk.fi for the host or Web server IP address.

There are two main types of DNS activities: lookups and zone transfers. Lookups happen when a DNS client, or a DNS server acting on behalf of a client, queries a DNS server for information. Typically lookups involve finding the IP address for a given hostname, the hostname for a given IP address, the name server responsible for a given domain, or the mail server for a given host.

Zone transfers happen when a DNS server requests all records pertaining to a part of the DNS naming hierarchy (the zone) from another DNS server. The requesting DNS server is called the secondary server and the serving one is the primary server. Zone transfers are expected to happen only among servers that should be replicated. Since DNS knows the details of how a network is structured (the names and IP addresses), this information may need to be protected.

2.4 Addressing

The Internet is based on hierarchical routing, which is reflected in its addressing system. The network addresses are divided into two parts, namely the network and host parts. The former defines the part of the network topology responsible for that address space, and the latter part defines the host. IPv4 has 32-bit addresses and the newer IPv6 extends this to 128 bits, which is expected to be sufficient for current needs. In both IPv4 and IPv6 the addressing space is divided into variable size prefixes.

Originally, there were three prefix classes of A, B, and C corresponding to 8, 16, and 24 bits for the network part in an address. The limitation of this model was that each prefix appeared with host addresses included in global routing tables, resulting in scalability challenges. As a result of a growth crisis, the *classless interdomain routing (CIDR)* was designed and deployed. CIDR supports provider aggregated addresses by allowing variable length network part in an address. This allows better utilization of the existing address spaces, especially class B networks and aggregate routing table entries. CIDR has significantly reduced the global routing tables, and it is used in IPv4 and IPv6 [1].

2.5 Routing

In this section, we briefly outline the basic routing process and then examine interdomain routing. We briefly present the *border gateway protocol (BGP)*, examine some of the current challenges for BGP, and finally consider compact routing, which is a family of routing schemes that aim for scalability.

2.5.1 Overview

Routing in a static network is straightforward, having each router determine directions for each possible destination. Routing in dynamic networks is more challenging, because the routing tables change and routing instructions need to be computed at runtime. The key question is where is the state and how often does it need to be updated?

The common approach is to broadcast routing state to all routers, which is exemplified in link-state routing protocols that broadcast link-state updates that are used to compute shortest path distances. To avoid excessive flooding of link-state updates, the common solution is to divide the network into routing domains and use this hierarchy to limit the propagation of link-state updates. Areas are extensively used in the OSPF, in which they are a network-dimensioning instrument. Hierarchies naturally occur in the interdomain context, in which autonomous systems reflect administrative boundaries.

A routing process is responsible for computing the forwarding table of a node. The routing process estimates the costs of incident links and communicates with its neighbors via these links. A routing algorithm is the mechanism that defines what information is exchanged with neighbors and how the forwarding tables are computed. The central purpose of a routing algorithm is to maintain a forwarding configuration in which nodes are mutually reachable by forwarding. It is often also desirable for the paths taken by forwarded packets to be optimal or near-optimal [197].

The Internet is based on hierarchical routing. The seminal work by Kleinrock and Kamoun published in 1977 showed how hierarchical clustering can be used to produce scalable routing tables [187]. The key idea is to cluster nearby nodes together and then combine clusters into superclusters, and continue this in a bottom-up hierarchical manner. As a result, unnecessary topological information gets abstracted from the routing tables, and the network scales well. Hierarchical routing results in routing table sizes on the order of $\sqrt{n}$. Hierarchical routing is used today by a variety of protocols in both interdomain (BGP, CIDR) and intradomain routing (OSPF).

2.5.2   Interdomain

The interdomain structure has resulted from developments in both technology and business models. It is a mixture of technological advances and business decisions driven by investors and the stock market. A current trend has been toward massively popular content services on the Internet. This has created pressure toward better network support of data delivery and dissemination. The need to be able to deliver vast amounts of data in an efficient and low-cost manner has given birth to CDNs and various peer-to-peer networks, such as BitTorrent networks.

CDNs charge for the data delivery service and are typically based on proprietary, closed solutions. BitTorrent and peer-to-peer networks, however, rely on peer-assisted data exchange. The latter rely on low-cost, mostly flat rate, connections between end-users and their providers. This new network behavior has resulted in various anti-peer-to-peer measures by Internet providers partly due to the fact that many P2P protocols, such as BitTorrent, do not take interdomain policies into account and thus are not ISP friendly.

The core Internet architecture was not designed to serve as critical communication infrastructure for society. Therefore, the economical and political context must also be analyzed and understood. The current question is whether viable economic models exist for Internet service provision. Business modeling is complicated by regulatory background, which varies by country. Telephone-carrier-based ISPs have been asking regulators for the ability to charge differentially, based on the application and content of traffic. This kind of discriminatory pricing may pose fundamental limitations for end users and limit their freedom.

Figure 2.4 illustrates interdomain routing with a number of autonomous systems. Overlay networks are implemented on top of the network layer topology as illustrated in the figure. The current interdomain practice is based on three tiers, namely tiers 1, 2, and 3. Tier-1 is an IP network that connects to the entire Internet using settlement-free peering. There are a small number of tier-1 networks that typically seek to protect their tier-1 status. A tier-2 network is a network that peers with some networks but relies on tier-1 for some connectivity, for which it pays settlements. A tier-3 network is a network that only purchases transit from other networks.

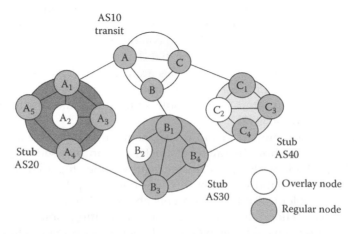

FIGURE 2.4
Example of interdomain routing.

> The three main AS categories are as follows [143]: *customer-to-provider (C2P)*, *peer-to-peer (P2P)*, and *sibling-to-sibling (S2S)*. In the C2P, a customer AS pays a provider AS for any traffic sent between the two. In the P2P category, two domains can freely exchange traffic between themselves and their customers but do not exchange traffic from or to their providers or other peers. In the S2S category, two domains are part of the same organization and can freely exchange traffic between their providers, customers, peers, or other siblings.

Gao's work formulated the AS relationships inference problem. Gao assumed that every BGP path must comply with the following hierarchical pattern: an uphill segment of zero or more C2P or S2S links, followed by zero or one P2P links, followed by a downhill segment of zero or more P2C or S2S links. Paths with this hierarchical structure are valley-free or valid. Paths that do not follow this hierarchical structure are called invalid and may result from BGP misconfiguration or from BGP policies that are more complex and do not distinctly fall into the C2P/P2P/S2S classification [143]. According to recent measurements, BGP tables miss up to 86.2% of the true AS adjacencies. The majority of these links are of the P2P type. This means that peering links are likely to be more dominant than have been previously reported or conjectured.

2.5.3 Border Gateway Protocol

The *border gateway protocol (BGP)* is responsible for connecting the different autonomous systems together, and it is the key protocol for building and maintaining the global routing table at interdomain routers. The current version of BGP is 4, and it incorporates support for CIDR and route aggregation to improve scalability (RFC 4271).

BGP is realized as a manually configured overlay network that uses TCP connections between peers. Routing updates propagate from peer-to-peer, and after receiving updates a BGP router updates its interdomain routing table based on the new information (the received path vectors).

BGP keeps a table of IP networks that are reachable either through peering links or transit links. Each IP address, or prefix, is associated with a vector of AS numbers that indicates the ASes that need to be traversed to reach the destination prefix. BGP is described as a *path vector* protocol, since it is built on this notion of a vector of AS identifiers. Moreover, BGP does not use intradomain metrics such as latency to make routing decisions; instead it uses network policies and rule sets to decide what paths are used in routing and forwarding.

2.5.4 Current Challenges

As a central component of the Internet, BGP is at the heart of the network and thus faces increasing scalability challenges as the global network grows. BGP scalability concerns stem from the observation that each interdomain router is expected to maintain routing paths to all valid network prefixes. Currently, there are almost 3×10^5 prefixes [1] in the global routing table, and this number is expected to grow in the near future because of site multihoming and provider-independent addressing. In addition to the space requirements, routing table updates poses several challenges. One is the frequency in which changes are propagated in the global backbone. Another concern is routing update oscillation that may result from router misconfiguration.

One way to alleviate BGP scalability concerns is to separate path selection from packet forwarding. This is exemplified in the *NIRA (a new interdomain routing architecture)* proposal that empowers users with the ability to choose a provider and domain level end-to-end path [354]. The motivation for this is that only users know when a path works or not. This model creates competition between paths that different ISPs offer, because users can choose the most suitable paths. In this model, the network comprises three parts for each sender and receiver—namely, the core region (tier-1), the uphill region that covers all possible paths from the sender to the core, and the downhill region covering all possible paths from the core down to the receiver. Each region can have its own routing protocols.

Another recent proposal, the *accountable internet protocol (AIP)*, replaces the subnet prefix in IP packets with a self-certifying autonomous system identifier and a suffix that is a self-certifying host identifier [9]. The key idea is to support domain-level routing instead of the current prefix-based routing. The motivation is that there are fewer autonomous systems than network prefixes. The proposal also combines domain-level routing with security by using self-certified identifiers that make it easier to make network entities accountable. The host identifiers are expected to be unique, which would support host mobility and multi-homing in a seamless way.

2.5.5 Compact Routing

As mentioned above, BGP faces significant scalability challenges, and recent measurements indicate that both the size of routing tables and the communication cost are increasing exponentially [190]. Prefix optimization techniques, such as CIDR, do not appear to be the most efficient solutions in the long run since they offer only a constant reduction in routing table sizes and they do not change the scaling behaviour of the network.

Compact routing has been proposed as a candidate solution for decreasing routing table sizes and improving network scalability. A routing scheme is said to be compact when it results in logarithmic address and header sizes, sublinear routing table sizes, and a stretch bounded by a constant. The compact routing schemes can be divided into two categories, specialized and universal. The former works only on some specific graphs, and the latter works on all graphs.

It has been shown that the classic link state, distance vector, and path vector routing algorithms exhibit routing table sizes on the order of $n \log(n)$ [144] with stretch-1 (the worst-case path length versus the shortest path). Moreover, hierarchical routing performs well only for graphs where large distances between nodes dominate. A universal stretch-1 compact routing algorithm has also $\Omega(n \log(n))$ [144]. One interpretation of this result is that shortest-path routing is incompressible, and to obtain smaller routing tables the stretch must be allowed to increase above 1. The Cowen and the Thorup-Zwick are two well-known nonhierarchical stretch-3 compact routing schemes. These name-dependent schemes utilize a set of landmarks to constrain updates and keep routing table sizes minimal. A routing table consists of entries for the shortest paths to all landmarks and nodes in the local cluster [144].

2.6 Multicast

Unicast is the dominant communication model for Internet applications. Multicast is the process of sending data from typically one sender to multiple receivers. This typically involves the creation of a multicast tree that is either source specific or shared by the communicating entities.

In general, the creation of an optimal multicast tree is equivalent to the *Steiner tree problem* that is known to be NP complete. This problem bears semblance to the *minimum spanning tree problem*; however, it considers only how to reach a specific subset of the nodes [348].

The multicast function can be implemented in the network level or it can be implemented in the application layer. Network-level multicast complements unicast as a basic networking primitive. Application-layer multicast, on the other hand, typically utilizes unicast. In this section, we first examine IP multicast and then consider overlay multicast techniques.

2.6.1 Network-layer Multicast

Multicast is essentially a one-to-many data delivery mechanism. Network-layer (or IP) multicast provides the multicast capability in the form of special multicast address ranges that are used by network routers to connect senders and receivers. Multicast differs significantly from unicast in that it decouples the senders and receivers. Moreover, since there may be a number of receivers for a multicast data packet, the network can optimize the transmission by replicating packets at the last possible moment in the network.

> *IP multicast* is a simple, scalable, and efficient mechanism to realize simple group-based communication. IP multicast routes IP packets from one sender to multiple receivers. Participants join and leave the group by sending a packet using the IGMP (RFC 1112) protocol to a well-known group multicast address.

The key components of IP multicast are

- IP multicast group address
- A multicast distribution tree maintained by routers
- Receiver driven tree creation

In order to receive multicast packets, receivers join a specific IP multicast group. A multicast distribution tree is constructed and maintained by routers for the group. All packets sent to the multicast IP address are then delivered by the multicast protocol to all receivers that have joined the group.

A multicast protocol is responsible for maintaining multicast trees that connect the members of multicast groups. There are two main categories of multicast algorithms, namely source-based trees and shared trees. The former is rooted at the router serving the source of multicast packets. This means that a tree is needed for each source; however, the trees can be optimal in terms of some metric. The latter is rooted at a specific router, called a *rendezvous point (RP)* or a *core*, that is responsible for maintaining the tree. In this case, the source sends data packets to the RP, which then is responsible for disseminating the data using the tree. The RP can then perform pruning operation to the tree to optimize the traffic.

Internet group management protocol (IGMP) is a protocol designed to allow the management of IP multicast groups memberships. IGMP is used by IP hosts and adjacent multicast routers to establish and maintain multicast groups. According to RFC 3171, addresses 224.0.0.0 to 239.255.255.255 are designated as multicast addresses. IGMP is based on UDP that is the common low-level protocol for multicast addressing. IP multicast, as IP in general, is not reliable, and messages may be lost or delivered out of sequence.

There are many different IP multicast protocols. The *protocol-independent multicast (PIM)* is a frequently used protocol that supports several different operating modes, namely *sparse mode, dense mode, source-specific mode,* and *bidirectional mode.* Several reliable multicast protocols have been developed—for example, the *pragmatic general multicast (PGM)* that extends IP multicast with loss detection and retransmission.

IP multicast groups are not very expressive. They partition the IP datagram address-space, and each datagram belongs at most to one group. Moreover, IP multicast is a best-effort unreliable service, and for many applications a reliable transport service is needed.

Multicast works well in closed networks; however, in large public networks multicast or broadcast may not be practical. In these environments universally adopted standards such as TCP/IP and HTTP may be better choices for all communication [168].

2.6.2 Application-layer Multicast

Given that IPv4 is still the prevailing network layer protocol and that it does not offer a native multicast mechanism, it is common to implement multicast on top of the TCP/IP protocol stack in the form of application-layer (or overlay) multicast. IP multicast requires routers to maintain per-group state or per-source state for each multicast group. A routing table entry is needed for each unique multicast group address, and the multicast addresses are not easily aggregated. Moreover, IP multicast still requires additional reliability and congestion control solutions.

Therefore, there is motivation for developing and deploying overlay multicast solutions. Indeed, many of the systems discussed later in this book are examples of these. In this section, we briefly outline the key motivation for application-layer multicast and the differences to network-layer multicast.

> An application-layer multicast system typically uses unicast communication between nodes to realize one-to-many communications. Data packets are replicated by the end hosts. These protocols may not be as efficient as IP multicast, because data may be sent multiple times over the same link. As an example, in a previous version of the Gnutella P2P protocol, one link was observed to be utilized six times for the same data [273]. This means that nodes establish communications either using UDP or TCP and forward messages using these links. The multicast tree construction algorithm is typically distributed and can take various metrics into account.

Figure 2.5 compares IP multicast and overlay multicast in the following categories: deployment, structure, transport, scalability, congestion control, and efficiency [174]. In terms of deployment, IP multicast requires multicast-capable routers, whereas overlay multicast

	IP Multicast	**Overlay Multicast**
Deployment	Multicast-capable routers	Deployed over the Internet
Multicast structure	Tree, interior nodes are routers, leaves are hosts	Typically a tree, both interior nodes of the structure and leaves are hosts
Transport layer protocol	UDP	TCP or UDP
Scalability	Limited	High (depends on solution)
Congestion control/recovery	No	Various, can utilize unicast (TCP) for node-to-node reliability
Efficiency	High	Low (varies), can suffer from high stretch and unoptimal interdomain routing
Example protocols	Protocol-independent multicast (PIM), Core-based trees (CBT), etc.	BitTorrent variants, Scribe, SplitStream, OverCast, etc.

FIGURE 2.5
Comparison of IP and overlay multicast.

is based on hosts and can thus be deployed easily over the Internet. Both approaches are based on trees, with the difference being that in IP multicast hosts do not participate in the tree other than as leaves. As mentioned, IP multicast is not widely deployed and hence its scalability is limited. It is, however, efficient, whereas overlay solutions may not utilize optimal paths and may incur more overhead.

2.6.3 Chaining TCP Connections for Multicast

Intuition suggests that overlay multicast typically incurs a performance penalty over IP multicast because of factors such as link stress, stretch factor, and end host packet processing. For example, early versions of the Gnutella P2P protocol used TCP, but later versions replaced it with UDP for performance reasons. Chains of TCP connections can offer an opportunity to increase performance compared to direct unicast. This performance improvement comes from finding an alternative overlay path whose narrowest hop in the chain (as perceived by TCP) is wider than the default path used by IP [192].

The expected TCP throughput as a function of the per-hop loss rates and RTTs can be modeled using the following equation derived in [247]:

$$T = \frac{s}{rtt\left(\sqrt{\frac{2p}{3}} + \left(12\sqrt{\frac{3p}{8}}\right)p(1+32p^2)\right)} \approx \frac{\sqrt{1.5}}{rtt\sqrt{p}} \tag{2.1}$$

This provides an estimate of the expected throughput T of a TCP connection in bytes/sec as a function of the packet size s, the measured round-trip time rtt, and the steady state loss event rate p.

A given hop in a chain of TCP connections either has local network conditions that limit its rate to a value below that of the upstream connections or is already limited by the rate of the upstream connections. Following the methodology used in [361], the aggregate RTT is defined as the sum of rtt_i along the path and the aggregate loss rate is defined as $1 - \sum 1 - p_i$ (assuming uncorrelated losses).

$$T \approx \frac{\sqrt{1.5}}{\sum rtt_i \sqrt{1 - \Pi(1 - p_i)}} \tag{2.2}$$

2.7 Network Coordinates

The latency of network communications is an important metric for choosing routes and peers on the network. This raises the question of how accurately latency can be predicted without prior communication. Recent network measurement systems indicate that latency prediction is feasible based on synthetic network coordinates [91, 101, 320, 349]. A network coordinate system might be used to select from among a number of replicated servers to request a file.

Vivaldi is a distributed algorithm that assigns synthetic coordinates to Internet hosts. It uses the Euclidean distance between the coordinates of two hosts to predict the network latency between them. In this system, each node computes its coordinates by simulating its position in a network of physical springs. The system does not require fixed infrastructure, and a new host can compute useful coordinates after obtaining latency information from some other hosts [101].

2.7.1   Vivaldi Centralized Algorithm

When formulated as a centralized algorithm, the input to Vivaldi is a matrix of real network latencies M, such that M_{xy} is the latency between x and y. The output is a set of coordinates. Finding the best coordinates is equivalent to minimizing the error (E) between predicted distances and the supplied distances. Vivaldi uses a simple squared error function:

$$E = \sum_x \sum_y (M_{xy} - dist(x, y))^2, \tag{2.3}$$

where $dist(x, y)$ is the standard Euclidean distance between coordinates of x and y.

Vivaldi places a spring between each pair of nodes for which it knows the network latency, with the rest length set to that latency. The length of each spring is the distance between the current coordinates of the two nodes. The potential energy of a spring is proportional to the displacement from its rest length squared: this displacement is identical to the prediction error of the coordinates. Therefore, minimizing the potential energy of the spring system corresponds to minimizing the prediction error E.

Vivaldi simulates the physical spring system by running the system through a series of small time steps. At each time step, the force on each node is calculated and the node moves in the direction of that force. The node moves a distance proportional to the applied force and the size of the time step.

Each time a node moves it decreases the energy of the system; however, the energy of the system stored in the springs will typically never reach zero, since network latencies do not reflect a Euclidean space. Neither the spring relaxation nor some of the other solutions, such as the simplex algorithm, is guaranteed to find the global minimal solution. Simulating spring relaxation requires much less computation than more general optimization algorithms.

2.7.2   Vivaldi Distributed Algorithm

In the distributed version of Vivaldi, each node simulates a piece of the overall spring system. A node maintains an estimate of its own current coordinates, starting at the origin. Whenever two nodes communicate, the two nodes measure the latency between them and exchange their current synthetic coordinates. In RPC-based systems, this measurement can be accomplished by timing the RPC; in a stream-oriented system, the receiver might echo a timestamp.

Once a measurement is obtained, both nodes adjust their coordinates to reduce the mismatch between the measured latency and the coordinate distance. A node moves its coordinates toward a point p along the line between it and the other node. The point p is chosen to be the point that reduces the difference between the predicted and measured latency between the two nodes to zero. To avoid oscillation, a node moves its coordinates only a fraction δ toward p.

A node initializes δ to 1.0 when it starts and reduces it each time it updates its coordinates. Vivaldi starts with a large δ to allow a node to move quickly toward good coordinates and ends up with a small δ to avoid oscillation. If two nodes have the same coordinates (the origin, for instance), they each choose a random direction in which to move. Algorithm 2.1 illustrates the update procedure.

2.7.3   Applications

A modified chord DHT (presented in Chapter 5) uses network coordinates to efficiently build routing tables based on proximity so that lookups are likely to proceed to nearby

---

Algorithm 2.1 Pseudocode for the Vivaldi update procedure

Data: s_c is the other host's coordinates, s_l is the one-way latency to that host, the initial value of δ is 1.0.

Function: *update*(s_c, s_l)

```
/* Unit vector toward other host                        */
```
Vector $dir = s_c - my_c$

$dir = dir \, / \, \text{length}(dir)$
```
/* Distance from springs rest position                  */
```
$d = \text{dist}(s_c, my_c) - s_l$
```
/* Displacement from rest position                      */
```
Vector $x = dir * d$
```
/* Reduce δ at each sample                              */
```
$\delta - = 0.025$
```
/* Stop at 0.05                                         */
```
$\delta = max(0.05, \delta)$

$x = x * \delta$
```
/* Apply the force                                      */
```
$my_c = my_c + x$

---

nodes. A node receives a list of candidate nodes and selects the one that is closest in coordinate space as its routing table entry; coordinates allow the node to make this decision without probing each candidate.

The modified chord utilizes coordinates when performing an iterative lookup. When a node n_1 initiates a lookup and routes it through some node n_2, n_2 chooses a next hop that is close to n_1 based on Vivaldi coordinates. In an iterative lookup, n_1 sends an RPC to each intermediate node in the route, so proximity to n_1 is more important than proximity to n_2.

2.7.4 Triangle Inequality Violation

For a network coordinate system to work, it needs to properly reflect the latencies between network hosts. When neighbour or peer selection is based on brute-force network measurements, the quality of the selection cannot be affected by *triangle inequality violations (TIV)*; however, when the number of nodes grows, performing these brute-force measurements may not be feasible. Then it is preferable to use a delay measurement system such as network coordinates discussed above. The potential challenge in using these systems is the assumption on the delay space that the triangle equality holds [340].

> Any three nodes on the Internet A, B, and C form a triangle ABC. Edge AC is considered to cause a triangle inequality violation if $d(A, B) + d(B, C) < d(A, C)$, where $d(X, Y)$ is the measured delay between X and Y. The triangulation ratio of the violation caused by AC in triangle ABC is defined as $d(A, C) = (d(A, B) + d(B, C))$.

It has been demonstrated that TIVs can cause significant errors in latency estimation based on network coordinate systems. As a potential remedy, a TIV alert mechanism has been proposed that identifies edges with severe TIVs [340].

2.8   Network Metrics

In this section, we examine metrics that characterize various properties of networks. Our focus is, in particular, on metrics that are useful in the design and deployment of overlay networks. We have already touched this issue when discussing routing. First, we briefly consider routing algorithm invariants, which are crucial for ensuring that the algorithms perform according to the specifications. These invariants and properties do not assess how well the paths perform that a routing algorithm maintains in a routing table. Therefore a number of metrics are needed to understand the quality of the paths, the state of the routers and nodes, and the properties of the network. We elaborate on the following metrics: *shortest path, routing table size, path stretch, forwarding load, churn,* and several other metrics.

2.8.1   Routing Algorithm Invariants

The correctness and performance of a routing algorithm can be analyzed using a number of metrics. Typically it is expected that a routing algorithm satisfies certain invariant properties that must be satisfied at all times. The two key properties are *safety* and *liveness*. The former states that undesired effects do not occur; in other words, the algorithm works correctly. The latter states that the algorithm continues to work correctly—for example, it avoids deadlocks and loops. These properties can typically be proven for a given routing algorithm under certain assumptions.

Safety and liveness can also be specified in terms of *soundness* and *completeness* [197] for a routing configuration. A configuration is sound if it includes paths for all node pairs that are reachable (have a path) after the network becomes quiet. A degenerate form of this configuration is one in which all nodes are unreachable. Completeness is used to ensure that all paths in the network are included in the configuration. Together these properties say that all nodes are reachable through the routing and forwarding system; however, they do not determine how optimal the paths are. Therefore, additional metrics are needed to assess the quality of the paths.

2.8.2   Convergence

Soundness and completeness (or safety and liveness) do not consider how quickly the routing algorithm works or converges when the network changes. They only ensure that from the viewpoint of the system invariants, the operation is correct. Indeed, convergence cost and time is an important metric for different kinds of routing systems, including overlay algorithms.

The dynamics of peers joining and leaving an overlay system is called churn, and it is an inherent property of P2P systems. Peer participation is highly dynamic. Typically, a large part of the active peers are stable and the remaining peers change quickly [312]. This means that P2P overlay networks must be designed in such a way that they tolerate high levels of churn. Indeed, many of the algorithms presented in Chapter 6 tolerate churn.

2.8.3   Shortest Path

The goal of a routing algorithm is to find the shortest paths between two destinations, *A* and *B*, that are reachable through the network. In order to do this, we need to have a metric for calculating these shortest paths and then create routing tables that reflect the paths according to distance. OSPF is an example of an intradomain protocol that computes shortest paths using link state routing. On the other hand, BGP is an example of a policy-based routing that calculates shortest paths based on policies and AS hops instead of, say, delay.

In general, the shortest path length between two nodes A and B is the minimum number of edges needed to traverse to reach A from B. The average path length is the average of the shortest path lengths between any two nodes. The average path length is a metric of the number of hops to reach a node from a given source.

2.8.4 Routing Table Size and Stretch

We can observe two conflicting goals in the design of routing algorithms, namely that the network paths used by a given router should be as short as possible and, at the same time, the routing table should be as small as possible. The two key metrics are the optimality of the paths and the size of the routing tables.

> The efficiency of a routing algorithm is measured in terms of its stretch factor— that is, the maximum ratio between the length (or delay) of a route computed by the algorithm and that of a shortest path (or delay) connecting the same pair of nodes [251].

Stretch signifies the degree of achieved performance in terms of the optimal choice. For overlay systems, there is an inherent overhead compared to IP routing with the benefit of deployability and scalability. The treatment for overlay multicast is a bit more challenging. Typically, the benchmark IP multicast tree would be assumed to consist of the optimal unicast paths.

We can extend the notion of a stretch to a multicast overlay tree as follows.

> Stretch for a multicast overlay tree is the ratio of the number of network-layer hops (or delay) in the path from the sender to a receiver in the multicast overlay tree, and the number of hops (or delay) required by the shortest unicast path between these two nodes, averaged over all trees and paths.

In addition to stretch, we have the routing table size as the other important metric. The routing table should hold only a fraction of nodes in the network, and the routing algorithm should not require global information about the nodes. For overlay networks, the aim is to support routing tables that have sublinear sizes to the number of nodes in the network (and the number of items in the network). The routing table data structure should also be efficiently realized using hardware and software.

These two metrics are in conflict, and a routing algorithm needs to balance between the size of the routing table and the optimality of the paths.

2.8.5 Forwarding Load

Another important metric is the forwarding load placed on routers in terms of packets, connections, and messages. For an IP router, forwarding load is measured in terms of incoming packets and the incurring per-packet delay. IP routers use hardware or software routing tables to look up destination interfaces for a packet given the packet's destination prefix. If a router cannot handle all incoming packets, its queues will become full and it will start to drop packets. This congestion is then handled at the edge of the network, following the end-to-end principle, and congestion avoidance is implemented in transport layer protocols, exemplified by the TCP congestion control algorithm.

Router forwarding load therefore is handled mostly at the edge for TCP/IP; however, overlay nodes are typically end hosts themselves, which makes stress an issue that has to be taken into consideration when designing an overlay algorithm. For an overlay node, forwarding load can be viewed to be the amount of traffic the node is processing at a particular time or time interval. This traffic has many components, namely control traffic pertaining to how the overlay network is structured (neighbors, super nodes, etc.) and the actual content.

In a multicast system, forwarding load can be expressed in terms of the branching factor (or replication factor) of each node. For overlay multicast systems, the load incurred from multicast forwarding compared to network level forwarding can be defined to be the number of identical packets sent by a node. For network layer multicast there is no redundant packet replication; however, an overlay multicast scheme may result in a number of unnecessary packet replications (called false positives).

2.8.6 Churn

Churn is a metric that is especially pertinent for P2P overlay systems. Churn pertains to the rate of arrivals and departures in the system. Typically, a large part of the active peers are stable and the remaining peers come and go. P2P overlay networks must be designed in such a way that they tolerate high levels of churn. Recent analysis of churn indicates that, overall, its characteristics are remarkably similar across different systems [271, 312].

> Churn is an inherent property of P2P systems and describes the dynamics of peer arrival and departure. High churn means that the system is highly dynamic, with peers coming and going.

Two metrics have been commonly used for churn in file-sharing systems, namely a node's session time and lifetime. The session time is the duration between the node joining the network and then subsequently leaving it. The lifetime is the time between when the node first entered the network and then left the network permanently. These two metrics are depicted by Figure 2.6. The availability of a node can be defined to be the sum of a node's session times divided by its lifetime. In one study, it has been argued that the session times of nodes in a DHT are more relevant than their lifetimes [271].

2.8.7 Other Metrics

Other important metrics that characterize a network include:

- Network diameter, which is the average minimum distance between any pair of nodes.
- Node degree, which is the number of links that the node has to other nodes in an undirected graph. This degree distribution is connected with the robustness of the network to node failures.

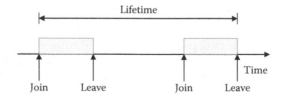

FIGURE 2.6
Session time in Churn.

- Locality-awareness and the properties of data, which are important for data lookup overlays and CDNs.
- Policy compliancy, which is important for routing that takes place across organization boundaries. BGP is the classic example of a policy-based routing protocol.

3

Properties of Networks and Data

This chapter examines the salient properties of networks and data communicated over the networks. We start with a characterization of data on the current Internet and discuss the growth rate of the global network. Both geographical and logical distribution of data are crucial when creating overlay networks over the Internet that ensure efficient data availability. We discuss the role of power-laws and small-worlds in networking.

In order to engineer efficient overlay systems, a lot of information is needed pertaining to the underlying network, the nodes and their characteristics, and the properties of the data that they subscribe, publish, and seek. This calls for various models, including models of the actual traffic distributions on the Internet (including their spatial and temporal characteristics), models of host connectivity, models of the dynamics of churn, and so on. In this chapter, we outline some of the fundamental characteristics of overlay networks.

3.1 Data on the Internet

> We are currently in the era of the exabyte in terms of annual IP traffic [78] and entering the era of the zettabyte (10^{21} bytes). Cisco's latest traffic forecast for 2009–2013 indicates that annual global IP traffic will reach 667 exabytes in 2013 [79]. The traffic is expected to increase some 40% each annum. Much of this increase comes from the delivery of video data in various forms.

Figure 3.1 presents Cisco's forecast estimates for monthly global IP traffic until 2011. According to these estimates, the Internet is growing fast. We can compare this estimate with the situation in 2005 when the global traffic was a bit over 2000 petabytes per month. The forecast predicts approximately eightfold increase in monthly traffic volume.

Figure 3.2 compares monthly traffic estimates for a number of content providers. The growth of data-intensive services is evident in the amount of traffic transmitted per month. We observe that Google and YouTube have by far the greatest bandwidth requirements. The estimates for US traffic for these two services in mid-2007 far surpassed the US Internet backbone at year end in 1998; in fact the traffic was over seven times larger. This gives an idea of the radical growth of the Internet in the last 10 years.

3.1.1 Video Delivery

Video delivery on the Internet is anticipated to see a huge increase, and the volume of video delivery is expected to be 700 times the capacity of the US Internet backbone in 2000. Cisco's study anticipates that video traffic will account for 91% of all consumer traffic in 2013.

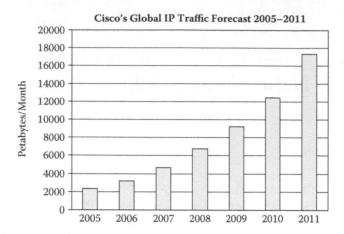

FIGURE 3.1
Cisco's global IP traffic forecast estimates 2005–2011.

The increasing video-related traffic creates a number of challenges for network engineering. Video files are typically large, and with the advent of high-definition content they will be even larger. This means that even a small adoption of a video delivery technology can result in significant shifts in traffic patterns. This unpredictability of traffic patterns makes network provisioning more difficult and may result in decreased quality of service for customers [145].

Flash crowds contribute to the unpredictability of the network. A flash crowd happens when a certain video or Web site becomes, typically unexpectedly, massively popular [119]. Flash crowds can be alleviated by using content replication and caching schemes. Another frequently used technique by Web sites is to detect unexpected demand for content and simplify the Web content to make it smaller.

Video delivery poses new challenges and opportunities for Internet service providers. Video consumes bandwidth, and with the emergence of flat rates consumers do not pay per megabyte. Moreover, the content may come from anywhere on the Internet, which may result in increased interdomain traffic charges for the ISP. This means that service revenue is no longer related to the connectivity revenue.

	Terabytes per month
BBC (UK) April 2007	196
Yahoo (UK) April 2007	353
Time Warner (US) May 2007	1129
ABC, NBC, ESPN, Disney (US) May 2007	1854
Yahoo (US) May 2007	2361
iTunes audio and video downloads (2006)	3500
Google (UK) April 2007	3709
MySpace (US) May 2007	4148
US Internet backbone at year end 1998	**6000**
Google and YouTube (US) May 2007	10956
Google and YouTube (worldwide mid-2007 Cisco estimate)	45750

FIGURE 3.2
Monthly traffic estimates for content services.

3.1.2 P2P Traffic

According to the study, peer-to-peer traffic will continue to grow but become a smaller component of Internet traffic in terms of its current share. The current P2P systems in 2009 are transferring 3.3 exabytes of data per month. The recent study indicates that the P2P share of consumer Internet traffic will drop to 20% by 2013, down from the current 50% (at the end of 2008). Even though the P2P share may diminish, most video delivery solutions, accounting for much of the traffic increase, will utilize overlay technologies, which makes this area crucial for ensuring efficient and scalable services.

3.1.3 Trends in Networking

Figure 3.3 presents a number of significant trends in IP networking and outlines their challenges and potential solutions. Current trends include P2P, Internet broadcast, both Internet and commercial video-on-demand (VoD), and high-definition content.

P2P presents a number of challenges for IP networks because it increases traffic and utilizes upstream for data exchange. This changes the customary usage of the network in which downstream dominates the traffic model. Therefore, IP networks need to be provisioned in such a way that possible upstream bottlenecks are eliminated. Caching can be seen as a potential solution to P2P traffic. Indeed, many current P2P protocols are able to take network proximity into account so that data can be obtained from a nearby P2P node.

Internet broadcast pertains to the dissemination of large media files or streams. Flash crowds are challenging because they make it difficult to provision the network in such a way that it can handle the expected demand for the content. This can be alleviated by using P2P content distribution technologies and multicast technologies. Since there is no global IP multicast support available, network layer multicast needs to be used in specific networks, such as metropolitan area networks or wireless access networks.

Internet VoD is becoming increasingly popular, and thus the growth of the traffic is a challenge for the network. This mostly affects the metropolitan area networks and the core networks. The solutions include CDNs and increasing the network capacity. Data compression can also be used to reduce the size of the media files. VoD can be cached, which makes it easy to cache. Commercial VoD is typically delivered in the metropolitan

Trend	Challenges	Solutions
P2P	Growth in traffic, upstream bottlenecks	P2P caching
Internet Broadcast	Flash crowds	P2P content distribution, multicast technologies
Internet Video-on-Demand	Growth in traffic, especially metropolitan area and core	Content Delivery Networks (CDNs), increasing network capacity, compression
Commercial Video-on-Demand	Growth in traffic in the metropolitan area network	CDNs, increasing network capacity, compression
High-definition content	Access network IPTV bottleneck, growth in VoD traffic volume in the metropolitan area network	CDNs, increasing network capacity, compression

FIGURE 3.3
Trends, challenges, and potential solutions for IP traffic.

area network, which needs to be provisioned accordingly. The core network is not burdened much by commercial VoD, because the content can be replicated to relevant MAN networks.

High-definition content also poses challenges, because due to higher quality the amount of data that needs to be transferred grows radically. Access networks are constrained by their IP television (IPTV) solution. CDNs and increasing the network capacity as well as compression are potential remedies.

3.2   Zipf's Law

> A power-law implies that small occurrences are extremely common, whereas large instances are extremely rare. This regularity or law is also referred to as Zipf or Pareto. Zipf's law is interesting for networked systems, because it has been shown that many different activities follow this law—for example, query distributions and Web site popularity. The linguist George Zipf first proposed the law in 1935 in the context of word frequencies in languages. For Web sites, the Zipf law means that large sites get disproportionately more traffic than smaller sites.

In this section, we give an overview of the Zipf distribution and two related distributions, namely Pareto and power-law distributions. Then we briefly discuss the implications for the Internet and P2P.

3.2.1   Overview

The Zipf distribution is concerned with the ranking of objects based on their popularity. The ranking is done by assigning the most popular object the rank of one, the second most popular object a rank of two, and so on. Zipf's law states that if objects are ranked according to the frequency of occurrence, the frequency of occurrence F is related to the rank of the object R according to the relation

$$F \sim R^{-\beta}, \tag{3.1}$$

where the constant is close to one.

The simplest verification of the applicability of Zipf's law is to plot the rank-ordered list of objects versus the frequency of the object on a log-log scale. On a log-log scale, the observance of a straight line is indicative of the applicability of Zipf's law. The Zipf distribution and power-law distributions are directly related, and they are different ways of looking at the same phenomena [5]. Zipf is used to model the rank distributions and power-law for frequency distributions.

The Zipf distribution is related to the Pareto distribution. Pareto was interested in the distribution of income, with the question of how many people have an income greater than x. Pareto's law is defined in terms of the *cumulative distribution function (CDF)*. The Pareto distribution gives the probability that a person's income is greater than or equal to x:

$$P[X > x] \sim x^{-k}. \tag{3.2}$$

A power-law distribution in its typical usage tells the number of people whose income is exactly x rather than how many people had an income greater than x. It is the *probability distribution function (PDF)* associated with the CDF given by Pareto's law

$$P[X = x] \sim x^{-(k+1)}m \tag{3.3}$$

where k is the Pareto distribution shape parameter.

3.2.2 Zipf's Law and the Internet

The ubiquitous nature of the Zipf distribution on the Internet can be explained using a growth model that is based on preferential attachment. We return to this topic in the section on scale-free networks. One key observation is that the multiplicative stochastic growth process results in a lognormal distribution in the number of pages of a Web site. When exponential growth of the Web is taken into account, the result is a power-law distribution (exponentially weighted mixture of lognormal distributions) [6].

Zipf's law has been used to model Web links and media file references. It therefore has profound implications for content delivery on the Internet. Efficient caching relies heavily on Zipf's law to replicate a small number of immensely popular files near the users. The distribution of the number of connections a host has to other hosts on the Internet has been shown to follow the Zipf distribution.

Given that video delivery is becoming increasingly popular, we can ask whether or not the Zipf distribution can be used to also model video delivery on the Internet. Sripanidkulchai et al. analyzed a workload of live media streams collected from a large CDN in 2004. They observe that on-demand streaming media popularity follows a two-sided Zipf distribution [301]. This distribution has a shallowed exponent for the high-degree nodes and a steeper exponent for the lower-degree nodes. They also observed exponentially distributed client arrival times within small time windows and heavy-tailed session durations.

3.2.3 Implications for P2P

The most straightforward approach to locate data in a P2P network is to flood queries (broadcast). Without a central server, all nodes flood queries to other peers, and as a consequence the network may become congested. Given the observation that Web resource and on-demand media distributions follow a one-sided or two-sided Zipf distribution, an interesting question is whether the P2P network (or any network in general) can take this data distribution into account in order to be more efficient. Indeed, recent results indicate that two-tier P2P systems such as Gnutella and Freenet, and structured overlays (DHTs), perform significantly better than simple flooding.

One defining difference in P2P file sharing and Web traffic is that most files that are shared in P2P networks are immutable, whereas Web pages are often updated and therefore mutable. Caching works much better for immutable resources than mutable ones, because immutable resources do not require a mechanism for coordinating updates (although versioning may be needed). Experimental results with the P2P system KaZaA indicate that clients typically fetch a file only once from the P2P network, whereas Web pages are more frequently requested by browsers. As a consequence, the KaZaA file popularity distribution differs from the typical Zipf distribution for Web resources [153].

3.3 Scale-free Networks

The way network nodes are connected, both the physical links as well as logical links, is vitally important in order to understand many properties such as scalability, performance, and resilience. The nature of both physical Internet connections and logical connections bears crucial importance to realistic topology generation and network simulation. In order to build good overlay designs, the underlying network must be understood and taken into account. This means that issues pertaining to network economics also need to be considered.

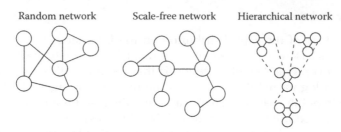

FIGURE 3.4
Three different types of networks.

Figure 3.4 illustrates different kinds of network types, namely random graphs, scale-free networks, and hierarchical networks. We briefly discuss these three types of networks. In many real systems it has to be assumed that clusters combine in an iterative manner, generating a hierarchical network.

> The Erdös-Rényi model of a random network starts with N nodes and connects each pair of nodes with probability p. The node degrees follow a Poisson distribution, and the tail of the degree distribution decreases exponentially. This indicates that most nodes have approximately the same number of links and that nodes that deviate from the average are very rare. The mean path length is proportional to the logarithm of the network size, $\log N$, which is indicative of the small-world property.

Recently it has been shown that many different self-organizing networks are scale-free. One of the distinguishing features of scale-free networks is preferential attachment. This results in large and busy hubs that route traffic and keep the diameter of the network small through multiple connections between different hubs. The hubs are useful in keeping the network connected and the diameter small; however, they are also a potential weak point of the network because of their central role in maintaining connectivity. Scale-free networks are characterized by a power-law degree distribution. Figure 3.5 gives an example of power-law distribution. In a power-law distribution, there are a few hubs that have many connections.

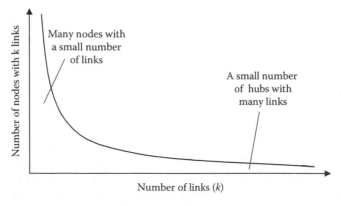

FIGURE 3.5
Power-law distribution.

Three central network characteristics have been described for scale-free networks based on analysis of complex networks:

1. Short average path length
2. High level of clustering
3. Power-law and exponential degree distribution (contrasting the Poisson degree distribution of the classic Erdös-Rényi model)

> The Barabási-Albert model supports incremental network growth of the scale-free topology. The model is based on three mechanisms that drive the evolution of graph structures, namely incremental growth, preferential connectivity, and rewiring.

A scale-free network can be grown by adding vertices to the graph one at a time and joining them to a fixed number m of earlier vertices. Each earlier vertex is chosen with probability proportional to its degree. It has been shown that the resulting graph has diameter $\log n$ for $m = 1$ and $\log n / \log \log n$ for $m = 2$ [34].

The World Wide Web graph (web pages are vertices, and hyperlinks are edges) and the bandwidth capacity of optical fibre connections between major US metropolitan areas have been demonstrated to exhibit scale-free properties. There are a few Web sites with a high number of links, which helps to further promote the popularity of the sites. It has been shown that the Internet follows power-law distributions both at the router level and AS level [126]. This means that the physical fabric of the Internet and the business interconnection of networks both can be considered to be scale-free networks. However, not all networks are scale-free and, for example, the Italian telephone system for out-going landline calls is more similar to exponential distribution than a power-law one.

Compact routing is a research area that investigates the limits of routing scalability [93, 190]. This research shows that shortest-path routing cannot guarantee routing table sizes that on all network topologies grow slower than linearly as functions of the network size. There exist static compact routing schemes designed for grids, trees, and Internet-like topologies that offer routing table sizes that scale logarithmically with the network size. Recent research indicates that logarithmic scaling on Internet-like topologies is fundamentally impossible in the presence of topology dynamics or topology-independent addressing.

3.4 Robustness

Given a certain expected network structure, a very interesting question is how easy it is to disrupt the network and partition it into disjoint parts. Cohen et al. [85] have shown analytically that networks in which the vertex connectivity follows a power-law distribution with an index of at most ($\alpha < 3$) are very robust in the face of random node breakdowns. A connected cluster of peers that spans the entire network can survive even in the presence of a large percentage p of random peer breakdowns. The following bound has been derived for p [284]:

$$p \leq 1 + \left(1 - m^{\alpha-2} K^{3-\alpha} \frac{\alpha - 2}{3 - \alpha}\right)^{-1}, \tag{3.4}$$

where m is the minimum node degree and K is the maximum node degree.

The Internet node connectivity has been shown to follow a power-law distribution with $\alpha = 2.5$ [85]. Similar investigation has been made for the Gnutella P2P network resulting

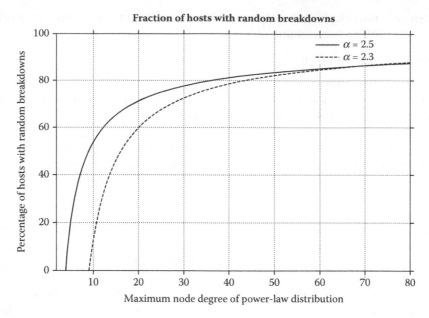

FIGURE 3.6
Resiliency of power-law networks.

in the observation that $\alpha = 2.3$ for Gnutella [284]. Figure 3.6 shows the above equation as a function of the maximum degree where the power-law parameter α was set to 2.3 and 2.5, respectively. Both the Internet and Gnutella present a highly robust topology. They are able to tolerate random node breakdowns.

For a maximum and fairly typical node degree of 20, the Gnutella overlay is partitioned into disjoint parts only when more than 60% of the nodes are down. Robustness is a highly desirable property in a network. The above equation is useful in understanding the robustness of power-law networks; however, it assumes that the node failures are random. Although a power-law network tolerates random node failures well, it is still vulnerable to selective attacks against nodes. Indeed, an orchestrated attack against hubs in the network may be very effective in partitioning the network.

3.5 Small Worlds

> Small-world networks are characterized by a graph degree power-law distribution. Most nodes have relatively few local connections to other nodes, but a significant small number of nodes have large wide-ranging sets of connections. The small-world topology enables efficient short paths because the well-connected nodes provide shortcuts across the network.

The notion of the small-world phenomenon can be traced back to the famous experiment by Stanley Milgram in the 1960s to assess people's social networks. His conclusions included that people were proficient at finding routes to other people even across continents.

In 1998 a certain category of random networks were identified to be small-world networks by Duncan Watts and Steven Strogatz [346]. This classification was based on two independent structural features: the clustering coefficient and the average distance between two nodes (average shortest path). Random graphs built using the classic Erdös-Rényi model feature small average shortest path with a small clustering coefficient. The small-world networks were observed to have this first property but to exhibit much larger clustering coefficient than would be expected.

Jon Kleinberg developed a mathematical model in 2000 for routing in small-world networks [186]. He investigated routing on lattices and showed that routing efficiency depends on a balance between the number of shortcut edges of different lengths with the coordinates in the lattice. More specifically, a d-dimensional lattice was used, with long-range links chosen at random according to the d-harmonic distribution. In this model, the probability of a random shortcut being a distance x away from the source is proportional to $1/x$ in one dimension, proportional to $1/x^2$ in two dimensions, and so on [346].

In a specific configuration in which the frequency of edges of different lengths decreases inverse proportionally to the length, simple greedy routing will find routes using only local information in $O(\log^2 n)$ hops on average, where n is the size of the graph [283]. More recent research shows that Kleinberg's analysis is tight and the algorithm achieves $\Theta(\log^2 n)$ delivery time. The expected diameter of the graph has also been shown to be $\Theta(\log n)$, a $\log n$ factor smaller than originally anticipated [222].

The small-world result is significant for overlay P2P networks since it allows retention of a small routing table with only a few distant contacts and still routes efficiently with only local information [220]. This result has been used in a number of P2P systems, notably Freenet [359] discussed in Chapter 4. Given the assumption that a network exhibits small-network properties, it should be possible to recover an embedded Kleinberg small-world network. This can be done by randomly selecting pairs of nodes to be included in a routing table and then potentially swapping them while minimizing the distances between a given node and its neighbors.

The Gnutella network has been observed to exhibit the clustering and short path lengths of a small world network. Its overlay dynamics lead to a biased connectivity among peers where each peer is more likely connected to peers with higher uptime [313]. Moreover, the Gnutella session lifetime has been observed to follow a power-law distribution with exponential cut-off. The session lifetime distribution of Gnutella might be an invariant characteristic independent of the protocol changes during the period 2002–2005 [163].

4

Unstructured Overlays

In this chapter, we examine a number of unstructured P2P overlay networks. Many of these solutions can be seen to be part of the first generation of P2P and overlay networks; however, they also can be combined with structured approaches to form hybrid solutions. We cover protocols such as Gnutella, BitTorrent, and Freenet and present a comparison of them. This chapter places special emphasis on BitTorrent, because it has become the most frequently used P2P protocol.

4.1 Overview

Overlay networks come in many shapes and forms. In general, there are two main classes of overlay and P2P networks: structured and unstructured. This chapter focuses on unstructured networks, in which there is no tight topology control by the algorithm. As mentioned in Chapter 1, unstructured P2P algorithms have been called *first generation* and the structured algorithms have been called *second generation*, respectively. They can also be combined to create *hybrid* systems [11, 213].

> Unstructured networks are typically based on random graphs following flat or hierarchical organization. Unstructured networks utilize flooding and similar opportunistic techniques, such as *random walks* [146, 225], *expanding-ring Time-to-Live (TTL)* search, in order to locate peers that have interesting data items.

Many unstructured P2P systems are based on keyword-based searching of interesting data items in the network. Keyword-based systems maintain an inverted list for each keyword that includes the identifiers for matching documents and the frequency of the keyword in each document. The P2P algorithm then distributes these lists. Search for multiple keywords involves intersection of the corresponding lists. The cost of maintaining and distributing keyword-based indexes can be alleviated by using more advanced data structures for list representation, such as probabilistic filters discussed in Chapter 7. Indeed, later versions of the P2P Gnutella protocol use Bloom filters for compact index representation.

Unstructured networks are contrasted by structured overlay networks, investigated in Chapter 5, which have stricter requirements on the topology and can facilitate more coordinated search efforts [313]. One benefit of unstructured networks stems from their opportunistic nature and that the topology makes little assumptions regarding the queries. Indeed, unstructured networks can support different kinds of query languages, whereas structured overlay networks typically require additional query-processing layers on top of the basic overlay.

4.2   Early Systems

The history of P2P systems is rooted in various client-server-based systems that introduced server-to-server communications in varying degrees. Internet was designed to support peer-to-peer systems, such as the *file transfer protocol (FTP)*, Telnet, *domain name system (DNS)*, and *unix-to-unix copy protocol (UUCP)*. The Internet architecture's emphasis on placing intelligence in the end hosts rather than the core routing infrastructure proved to be a fertile ground for developing different end-to-end interactions between hosts.

As a classical example, we can take the Usenet news systems, in which news servers exchange news articles in order to propagate them. Usenet is a client-server network from the viewpoint of the clients; however, it exhibits P2P communication from the viewpoint of servers. We can view the e-mail system and *simple mail transfer protocol (SMTP)* [261] from a similar angle. In SMTP, the relaying network of mail agents can be seen to be a P2P network. The Web introduces the capability for P2P, since any node can be a Web client or a Web server. These examples can be seen as unstructured overlays since they do not impose strict requirements on the connections between peers.

4.3   Locating Data

A significant feature of P2P systems is how they locate the desired data items in the distributed environment. The four common techniques are

- Central index: This model was used by Napster. When a node joins, it sends a list of locally available files to the index server. The index server then performs queries on behalf of the clients. When a query has matching files, the peer that sent the query receives a list of the peers that have the actual data file. Although the central index provides guarantees for completeness and can be used to ensure global end-to-end reachability, it is the weakest point in the system from the viewpoint of security. In addition, the central nature necessitates that an organization manage and maintain it.

- Flooding: In this model, there is no central index, but each peer maintains an index of files it is sharing with other peers. A query is then propagated in the network by peers, and peers that have matching data items send the result set directly to the node that initiated the query. This method avoids the single point of failure of Napster; however, it introduces significant overhead into the network due to the flooding technique.

   Flooding-based search techniques are typically effective in locating highly popular data items. They are also robust in high churn environments, in which peers join and leave the system. These techniques are not very good in locating rare data items. Moreover, in unstructured P2P systems, the peer load in many cases grows linearly or super-linearly with the total number of queries and system size.

- Heuristic key-based routing: Given the inherent performance and security issues in the basic flooding model, a number of heuristic key-based routing techniques have been developed. Freenet is a classic example of a P2P system using the key-based routing model. In the model, each file is associated with a key, and files with similar keys are clustered together. The motivation is that queries can then be forwarded to a cluster instead of flooding the network. The limitation of this heuristic approach is that the network does not guarantee that all matching data items are found.

- Structured models: As mentioned above, centralized, flooding, and heuristic key-based models have certain limitations. In order to overcome these limitations, many different structured P2P models have been proposed. The aim of these models is to be able to offer the efficiency and completeness of search results of Napster with the decentralization of Gnutella and Freenet. A large part of structured overlays has focused on DHTs, which are examined in Chapter 5. Although many DHTs can offer decentralized and scalable exact-match search, more complex query processing requires further solutions and is currently an active topic of research and development.

Users of P2P file sharing networks, such as Gnutella, face the question of whether or not to share resources to other peers in the community. They face essentially a social dilemma of balancing between common good and selfish goals. The selfish behavior often encountered in P2P networks in which peers only download files and do not make resources available on the network is called *free-riding*. Free-riding occurs because the peers have no incentives for uploading files. Free-riding becomes a major problem when significant numbers of peers consume network resources while not contributing to the network. In the context of P2P this is often referred to as *tragedy of the digital commons* [52, 167].

4.4 Napster

The era of mainstream P2P file sharing can be seen to have started from Napster, which was launched in 1999. Although relying on direct file exchanges between peers, Napster employed a centralized file index hosted by the Napster service [285]. In this model, each peer provides a file list to the centralized file search service that maintains the file index.

Figure 4.1 outlines the key components of Napster. The centralized directory maintains the file lists of the peers. Peers query the directory server in order to find other peers that host files that match the query (step 1 in the figure). Once there is a match for a peer's query, the server forwards the address of the peer that stores the data item that matched the query (step 2). Finally, the peer that issued the matching query can directly contact the peer that has the data (step 3).

The motivation for this model is that the transfer of the file lists does not require much bandwidth, allows easy management, and ensures that the index is complete (i.e., has all the files available in the system). On the other hand, this model has a single point of failure. The service becomes unavailable if the index is not working properly. Although Napster popularized P2P file sharing, it also created a number of issues pertaining to copyright issues, and the digital music sharing capability was ultimately closed due to a lawsuit filed by the Recording Industry Association of America (RIAA).

In addition to the many legal questions arising from P2P file sharing, the observation that centralized components create challenges for the growth of the P2P network has led to many advances in P2P networking. Gnutella [272] goes beyond Napster in that it is entirely decentralized, thus avoiding the limitation of the centralized index. Gnutella and many similar networks utilize the flooding model in which a file query is broadcast on the network and peers propagate the query. Each peer examines an incoming query and evaluates the query against a list of local files. Although this model results in a decentralized

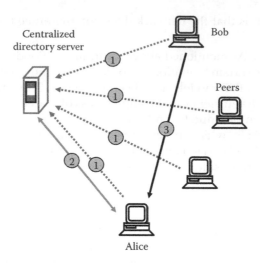

FIGURE 4.1
Overview of Napster.

system and a file index that is distributed over the peers, flooding has a lot of overhead in terms of network bandwidth and message processing, and it may take some time to find relevant files as the query is propagated in the network.

The realization that flooding is not efficient has led to a number of variants of the basic Gnutella protocol and, more recently, to a number of more sophisticated structured algorithms that impose constraints on the way peers are organized into the P2P network in order to optimize processing and find data more efficiently.

4.5   Gnutella

Gnutella is a classic example of a decentralized unstructured P2P network that relies on flooding to be able to find peers that have desired data items [70, 272].

> Gnutella is a decentralized P2P network that distributes both the search and download capabilities. The protocol uses flooding to find peers with matching data items and then uses direct file exchanges between peers. The protocol performance has been improved by introducing structure using hubs, called ultra nodes, and by employing proxy-based firewall traversal.

In this section, we discuss the classic version of the protocol and some later improvements. The Gnutella protocol is currently being developed by the Gnutella Developers Forum. There are a number of extensions available for the base protocol that include query routing enhancements, UDP-based querying and file transfers, XML metadata, parallel downloading, etc.

4.5.1   Overview

Figure 4.2 presents an overview of the classic Gnutella protocol. The two important parts of the protocol are

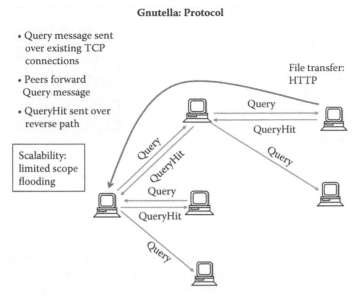

Gnutella: Protocol

- Query message sent over existing TCP connections
- Peers forward Query message
- QueryHit sent over reverse path

Scalability: limited scope flooding

File transfer: HTTP

Query
QueryHit
Query
QueryHit
Query
Query
QueryHit
Query

FIGURE 4.2
Overview of Gnutella.

- Search, which is about locating the peers that have desired data items.
- Download, which is the process of transferring the actual data items from the peers.

The bootstrapping of the P2P network by obtaining a set of known peers is not part of the Gnutella protocol. Typically, a set of predefined bootstrapping points are used. Once an address of at least one known peer has been obtained, the Gnutella client can try to connect to the P2P network using TCP/IP.

Searching is performed using the Query messages that are forwarded by the peers. Flooding is used to distribute queries over the overlay with a limited scope. Each peer maintains its own index of local data items. When a peer receives a query, it sends a list of data items that match the query to the peer that originally sent the query. The peer hosting the data sends the QueryHit message back to the origin peer using the reverse path. The peers then negotiate the file transfer and use HTTP to transfer the file.

The classic Gnutella protocol uses the following five message types:

- Ping. This message is used to discover peers on the network. A new peer sends a broadcast Ping message to announce its availability. The Ping message results in a corresponding Pong message that contains information about the peer that received the Ping message, such as network information and number of data items.
- Pong. The Pong message is sent as a reply to the Ping message.
- Query. The Query message is used to search for a data item (file). This message contains a search string. Each peer that receives the Query message checks the search string against its local database of file names. The Query message is propagated in the P2P network until the hop count reaches its maximum value.
- QueryHit. QueryHit is a reply to the Query message and contains information needed for downloading the file. The file transfer is first negotiated and then performed directly by the peers.
- Push. This message is a download request that is used by peers behind firewalls to trigger push-based file transfer.

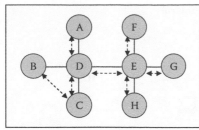 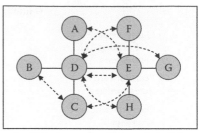

| Perfect mapping for message from A. Link D–E is traversed only once. | Inefficient mapping that results in link D–E being traversed six times |

FIGURE 4.3
Example of efficient and inefficient routing tables in Gnutella.

4.5.2   Searching the Network

In an early version of Gnutella (version 0.4), the number of actively connected nodes of a peer was relatively small, say five. When issuing a query, the peer would send the query message to each of these actively connected nodes, and they would then propagate the query. This was repeated until a predetermined number of hops, maximum of seven, was reached (the TTL).

Flooding works reasonably well for small- to medium-sized networks. The cost of searching in a Gnutella-style network using flooding increases super-linearly to the number of nodes in the system. Searching is roughly of exponential complexity, because t^d hops are involved where t is the time to live and d is the number of peers per node.

Figure 4.3 illustrates efficient and inefficient routing tables in terms of network proximity. In this example, a message will traverse the link D-E six times in an inefficient configuration. This illustrates the need to take network proximity into account [272].

More recent versions of Gnutella incorporate more structure in order to make the network more efficient and take locality better into account. Since version 0.6, Gnutella has been a composite network consisting of leaf nodes and ultra nodes. The leaf nodes have a small number of connections to ultra nodes, typically three. The ultra nodes are hubs of connectivity, each being connected to more than 32 other ultra nodes. Figure 4.4 illustrates this two-tier Gnutella architecture.

When a node with enough processing power joins the network, it becomes an ultra peer and establishes connections with other ultra nodes. This network between the ultra nodes is flat and unstructured. Then the ultra node must establish a minimum number of connections with client nodes in order to continue acting as an ultra node. These changes attempt to make the Gnutella network reflect the power-law distributions found in many natural systems. The maximum hop count was lowered to four to reflect this new structure.

In Gnutella terminology, the leaf nodes and ultra nodes use the *query routing protocol* to update routing tables, called *query routing table (QRT)*. The QRT consists of a table of hashed keywords that is sent by a leaf node to its ultra nodes. Ultra nodes merge the available QRT structures that they have received from the leaf nodes and exchange these merged tables with their neighboring ultra nodes.

Query routing is performed by hashing the search words and then testing whether or not the resulting hash value is present in the QRT of the present node. Ultra nodes perform this query test before forwarding a query to a leaf node or to a neighboring ultra node. The query ends when enough sources have been found (250 results).

If a match is found on a leaf node, this node contacts the peer that originated the query. The classic Gnutella protocol used reverse path routing to send a message back to this origin

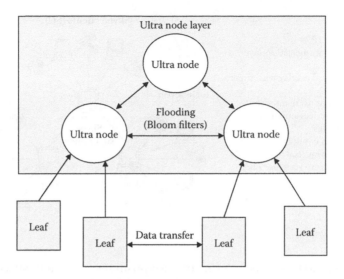

FIGURE 4.4
Two-tier Gnutella.

peer. Later incarnations of the protocol use UDP to directly contact the origin peer. As a result of this change, the Gnutella network is not burdened with this traffic.

When a file download is started, the peer that has the data and the origin peer negotiate the file transfer. If there is no firewall or NAT between the communications, the origin peer can contact the peer that has the data. Otherwise, the origin peer sends a special message called *push request* to the peer having the data. This message will trigger the other peer to initiate the connection and to push the file to the origin of the query. The historical Gnutella protocol used the same route for both queries and push messages; however, this was found to be unreliable due to the dynamic nature of the P2P network. As a solution, special entities called push proxies were introduced. Push proxies are announced in the search results, and they are commonly the ultra nodes of the leaf node in question. A push proxy sends a push request on behalf of a peer. The push proxies alleviate traffic concerns pertaining to push messages, and they offer more reliability, since ultra nodes are assumed to be more stable than leaf nodes.

When the Gnutella software instance is closed by a user, the software saves the list of leaf nodes and ultra nodes that it was actively connected to, as well as the peer addresses obtained from pong packets. This information is used to bootstrap the next Gnutella session.

A random walk scheme has been proposed as a replacement for the flooding algorithm. In this approach, each node chooses a random neighbor and sends the request only to this node. The random walk scheme was combined with proactive data replication, and they were found to improve the system performance [214] significantly.

4.5.3 Efficient Keyword Lists

The exchange of keyword lists and other metadata between peers is crucial for P2P networks. Ideally, the state should be such that it allows for accurate matching of queries and takes sublinear space (or near constant space). Gnutella minimizes the state needed for keywords by hashing them.

The later versions of the Gnutella protocol uses Bloom filters [32] to represent the keyword lists in an efficient manner. A Bloom filter is a probabilistic representation of a set that allows constant time membership tests. We return to Bloom filters and their variants in Chapter 7. Although Bloom filters require much less space than an ordinary keyword list,

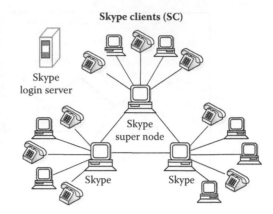

- Skype is P2P
- Proprietary application-layer protocol
- Hierarchical overlay with super nodes
- Index maps usernames to IP addresses; distributed over super nodes
- Peers with connectivity issues use NAT traversal or communicate via super node relays

FIGURE 4.5
Overview of Skype.

the price for the compactness is in terms of false positives. A false positive in this context means that the filter reports an unnecessary keyword match. The false positive probability can be tuned to a suitable level. In Gnutella, each leaf node sends its keyword Bloom filter to an ultra node, which can then produce a summary of all the filters from its leaves and then send this to its neighboring ultra nodes.

4.6   Skype

In addition to Napster, Skype[1] is another example of a proprietary P2P network for voice calls over the Internet. Skype has over 40 million active users, which makes it one of the most popular P2P networks in the world. P2P technology is used to make the system cost efficient and scalable. Skype allows people to call each other over the network using voice over IP free of charge and charges for calls to landlines and mobile phones. The system is based on the closed source Skype protocol. The user directory is decentralized and distributed among the nodes of the network; however, user authentication is done using a centralized server. The authentication servers are used to join the Skype network and obtain a list of so-called super nodes that are used to route calls. This cache of super nodes is periodically updated.

Figure 4.5 gives an overview of the Skype system. Some Skype peers that are publicly addressable, meaning that they are not behind NATs and firewalls, are super nodes that are used as rendezvous points for other users who are behind firewalls. Skype tries to utilize direct connections between clients by using STUN and TURN, but if this is not possible, it may need to use the super nodes for communication. Each Skype client maintains a host cache of known super nodes. The super nodes are used for locating other users and call routing. Unlike traditional landline calls, Skype encrypts all communications with a 128-bit cipher (AES) and uses RSA to transmit session keys.

4.7   BitTorrent

BitTorrent is currently the de facto P2P file-sharing protocol for distributing large amounts of data [103]. The protocol is very popular, and, according to some estimates, it accounts for

---

[1] www.skype.com

roughly 35% of all Internet traffic. The protocol was designed by Bram Cohen and released in 2001. The protocol specification is maintained by his company, BitTorrent, Inc.[2]

> BitTorrent is based on the notion of a *torrent*, which is a smallish file that contains metadata about a host, the *tracker*, that coordinates the file distribution and files that are shared. A peer that wishes to make data available must first find a tracker for the data, create a torrent, and then distribute the torrent file. Other peers can then use information contained in the torrent file to assist each other in downloading the file. The download is coordinated by the tracker, which is also the original source of the data (the *seed*). In BitTorrent terminology, peers that provide a complete file with all of its pieces are called *seeders*.

The efficiency and scalability of the BitTorrent protocol results from the simple requirement that each peer participating in the network share the data it has downloaded to others. This means that even a seed with slow network connection can distribute data in a scalable fashion, given that there are enough faster peers. After a peer has downloaded a file, it may choose to keep it available for others. In this case, the peer becomes a new seed for the file and thus improves the availability of the file. As more peers join the group of peers, called a *swarm*, the probability of finding a peer with the file or parts of it increases.

The simple nature of BitTorrent has resulted in numerous protocol implementations, and it is also easily deployable on the Internet due to its reliance on TCP. BitTorrent is very attractive for Internet hosting scenarios because it can help to reduce networking and hardware costs. A BitTorrent file download differs from an HTTP request in three basic ways:

- BitTorrent uses multiple parallel connections to improve download rates, whereas Web browsers typically use a single TCP socket to transfer HTTP requests and responses.
- BitTorrent is peer-assisted, whereas HTTP request is strictly client-server.
- BitTorrent uses the random or rarest-first mechanisms to ensure data availability, whereas HTTP is incremental.

> BitTorrent offers better resistance to *flash crowds* than a standard Web server because peers assist each other in downloading files. A flash crowd happens when a Web resource becomes overwhelmingly popular—for example, when the link is propagated in an epidemic fashion by other sites and media.

BitTorrent attempts to solve the *broadcasting problem*, which has the goal of disseminating M messages in a population of N nodes in the shortest time. In an environment in which the nodes have bidirectional communications and the same bandwidth, the lower bound on download time (rounds) is given by $M + \log_2 N$, and the unit is the time it takes for two nodes to exchange a message [239]. This problem can be solved optimally with a centralized scheduler; however, BitTorrent lacks this centralized component, and, furthermore, it does not have a completely connected graph. BitTorrent therefore has a heuristic approach to solving this problem that works very well in practice.

Figure 4.6 illustrates the BitTorrent protocol. The tracker identifies the swarm and helps the peers to trade the pieces. The tracker is a Web server accepting HTTP or HTTPS GET

---

[2] Specifications can be found at http://bittorrent.org/

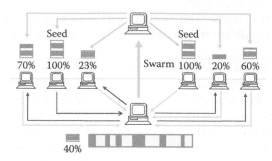

FIGURE 4.6
Overview of BitTorrent.

requests for information about a particular torrent. The tracker maintains state about the status of a torrent (for example, the peers and information about the pieces they have). The tracker also keeps overall statistics about the torrent.

Initially, the tracker identifies the initial seeds. When peers complete download and continue to share the file, they become peers as well. The file is downloaded in pieces, and, using the SHA-1 hash values for the pieces, a BitTorrent client can incrementally check the integrity of each downloaded piece. If a piece fails the authenticity test, it is dropped. In BitTorrent terminology, a piece refers to a part of the downloaded data that can be verified by a SHA-1 hash. A *block* is a part of data that a client may request from a peer. Each piece is made from two or more blocks.

Figure 4.7 presents the interactions in the BitTorrent protocol. In the first step, the seeder uploads a torrent file to a torrent server. As mentioned, this file contains information about the data, including the pieces and their hashes. The torrent file also identifies the tracker that coordinates the cooperative file sharing. The server makes the torrent available through various search techniques. In the second step, the seeder provides the torrent to the tracker, which then refers initial requests for pieces to the original seeder. In the third step, clients search for files and find the torrent file using the torrent server. The clients can then contact

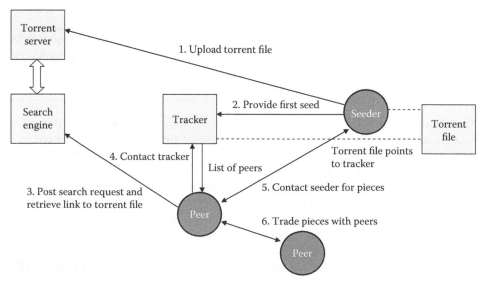

FIGURE 4.7
Key interactions in BitTorrent.

the tracker and participate in the P2P network (step 4). The initial seeder provides pieces that are not yet in wide circulation in the network (step 5). Eventually, the new client (and peer) will start to share pieces with other peers (step 6).

The two important characteristics of the BitTorrent protocol are *peer selection* and *piece selection*. The former is about selecting peers who are willing to share files back to the current peer. The latter is about supporting high piece diversity.

Tit for Tat in Peer Selection Based on Download Speed   This mechanism prefers peers who share files with the current peer based on the speed with which they are willing to share. The mechanism uses a choking/unchoking mechanism to control peer selection. The goal is to foster reciprocation and mitigate free riders.

> In game theory, tit for tat is an effective strategy for the iterated prisoner's dilemma. A player following this strategy will initially cooperate, then respond in kind to an opponent's previous move. This means that if the opponent was cooperative, then the player is cooperative. Otherwise, the player is not [16, 17]. The BitTorrent algorithm uses tit-for-tat strategy for peer selection.

In BitTorrent terminology, peer A is *interested* in peer B when B has pieces that A does not have. A is not interested in B's pieces if it already has them. Peer A is choked by peer B when B has decided not to send data to A. Peer A becomes unchoked by peer B when B is ready to send data to A. This does not necessarily mean that peer B is uploading data to A, but rather that B is willing to upload A if A issues a request.

The BitTorrent reference client software uses a mechanism called *optimistic unchoking*, in which the client uses a part of its available bandwidth for sending data to random peers. The motivation for this mechanism is to avoid a bootstrapping problem with the tit for tat selection process and ensure that new peers can join the swarm.

Local Rarest First for Piece Selection   Pieces are prioritized by their rarity in the local setting in order to enable high piece diversity. This is known as the local rarest-first algorithm because it bases the selection on the information available locally at each peer. Peers independently maintain a list of the pieces each of their remote peers has and build a rarest-pieces set containing the indices of the pieces with the least number of copies. This set is updated every time a remote peer announces that it acquired a new piece and is used by the local peer to select the next piece to download.

This policy is used except for the first four pieces, which are chosen using a random-first policy. This difference stems from the fact that when a new BitTorrent client joins the swarm, it does not have any pieces. Therefore it makes sense to randomly choose a piece from peers that unchoke it. The main goal in this phase is to simply obtain pieces so that trading can start. Since rare pieces are by definition less frequent than random pieces, the random selection is reasonable.

4.7.1 Torrents and Swarms

All users in a particular swarm are interested in obtaining the same file or set of files. In order to connect to a swarm, peers download the torrent file from the content provider. This is typically done using HTTP GET request. Peers communicate with the tracker when joining or leaving the swarm. They also communicate periodically as the download progresses, in the typical case every 15 minutes. The tracker is responsible for maintaining a list of currently active peers.

The metadata included in the torrent specifies the name and size of the file or files. Each file is divided into equal-sized pieces. These pieces are identified in the torrent file using SHA-1 hash fingerprints of the pieces. The piece sizes vary from 64 KB to 4 MB, the typical size being 256 KB. The SHA-1 hash fingerprints are used to check data integrity.

Typically the suffix *.torrent* is used to identify torrent files. These files are generally published on Web sites and registered with a known tracker. A torrent file can be expected to contain the following information:

- URL of the tracker
- File information (names, lengths, piece length)
- SHA-1 hash code for each piece

4.7.2 Networking

Assuming that the last-hop network connections are the limiting links, each peer provides an equal share of its available upload capacity to its active peer set. This sharing rate is determined by the upload capacity of a peer and the size of its active set. The official BitTorrent reference implementation sets the active set proportional to the square root of the upload capacity; however, this is set to a constant value in some other implementations.

The control traffic required for data exchange is small. Each peer transmits messages indicating the pieces they currently have and messages signaling their interest in the pieces of other peers.

BitTorrent utilizes multiple TCP connections in the data transmission. In order to minimize adverse effects due to competition between multiple TCP flows, the BitTorrent client uses at most five simultaneous TCP connections for uploading blocks. The download process can utilize more connections.

During the initial startup phase, the BitTorrent protocol creates new socket connections to peers until the number of connected peers has reached some preset maximum (typically 40). The protocol also accepts incoming connection requests from other peers until another maximum threshold value is reached (typically 80). The client software asks the tracker for a new list of peers periodically, say every 5 hours, or if the number of peer connections reaches a minimum value. After a connection is established, the peers exchange information to determine if the connection can be used to exchange data (interested) or, in the case that blocks cannot be exchanged, it is choked. For a client to be able to download data from a peer, the client needs to be interested and not choked by the peer.

4.7.3 Choking Mechanism

Choking pertains to connection management, and it is used to pause communications with peers. A choking algorithm should meet several requirements. It should be able to work well with the transport-layer protocol, namely TCP. In addition, it should avoid oscillations in connection management. BitTorrent uses a download rate–based tit-for-tat strategy to determine which peers to include in the current active set. Each round, a peer sends data to unchoked peers from which it downloaded most data in the recent past. This strategy aims to provide positive incentives for contributing to the system and inhibit free-riding.

Each BitTorrent peer always unchokes a fixed number of other peers (four is the default). TCP's built-in congestion control is used to saturate upload capacity. Therefore, current download rate determines which peers are unchoked. The reference implementation uses a moving 20-second average to calculate the download rate.

In order to allow new nodes to enter the P2P system, BitTorrent uses what is called optimistic unchoking, in which a small number of randomly chosen peers are unchoked.

Peers that do not upload data quickly enough gain reciprocation and are removed from the active tit-for-tat round and are returned to the choked status. To prevent oscillations between choked and unchoked peer states, BitTorrent peers recalculate the status every 10 seconds. This 10-second period is long enough for TCP to run the *slow start* algorithm and get flows to their full capacity.

The greedy approach of uploading to peers that provide the best download rate is limited by not being able to discover if some of the unused connections are faster than the currently used ones. To address this limitation, BitTorrent has a single optimistic unchoke active that is unchoked irrespective of the download rate. This optimistic unchoking is evaluated every 30 seconds.

Peers reciprocate uploading to peers that upload to them. The goal is to have several connections transferring files simultaneously in both directions and use trial and error to find unused capacity in the network.

The two key rules for the choking algorithm are

- A peer uploads to peers from which it is downloading with the fastest rate. This is reevaluated periodically (typically every 10 s).
- A peer optimistically unchokes peers. This is changed periodically (typically 30 s).

4.7.4 Antisnubbing

From time to time, a BitTorrent peer becomes choked by all peers from which it was receiving pieces. This means that the peer either has poor download capacity due to its limited upload capacity or it does not have the requested pieces. This situation is typically resolved by an eventual optimistic unchoke that finds a better peer than any of the current ones.

To work around this problem, if over a minute has passed for a particular peer connection without any download activity, the BitTorrent protocol assumes that the connection to the peer is *snubbed* and any upload activity is ceased, except in the case of an optimistic unchoke. Moreover, a BitTorrent client may optimistically unchoke more than one client when it is snubbed from all its peers.

4.7.5 End Game

The final stages of a file download require more examination. In BitTorrent swarms, it may happen that a client has to wait for the last few pieces to be downloaded before the download is complete. In order to complete the download in a swift manner, the protocol sends requests for all of its missing pieces to all of its peers. To avoid unnecessary transmissions of the pieces, the client keeps the other peers informed about its current download situation and sends cancel when a new piece is obtained. The BitTorrent specification does not explicitly define when the so-called *end mode* should be started. Some clients enter this end game mode when they have requested all pieces.

4.7.6 Trackerless Operation

BitTorrent can also be used in a trackerless setting, in which every peer implements the tracker functionality. This kind of decentralized behavior is achieved by utilizing a structured distributed as table (DHT) algorithm. A DHT algorithm known as the mainline DHT is used by many BitTorrent clients. This DHT algorithm is based on the Kademlia system discussed in Chapter 5 and uses UDP for communications.

BitTorrent clients include a DHT node, which is used to contact other nodes in the DHT to get the location of peers to download from using the BitTorrent protocol.

4.7.7 BitTorrent Vulnerabilities

So far we have not considered security of the BitTorrent protocol. In principle, most P2P networks do not provide guarantees on reliability and trustworthiness of the service. The huge popularity of BitTorrent makes it a possible instrument for all kinds of attacks against services and motivates the examination of the possible security threats and solutions. We briefly outline the key issues and return to them later in Chapter 9.

As the size of a swarm grows, so does the probability that it contains malicious peers. *Distributed denial of service (DDoS)* is a common form of attack that is launched against services, its aim being to overwhelm the network and server with a flood of packets. We observe that the BitTorrent protocol does not address these attacks.

There are also some exploits specifically targeted at BitTorrent that aim to increase the download performance of rogue peers. Rogue peers can download pieces only from seeds, which have negative consequences for the performance of the swarm since the upload capacity of the seed is used. Moreover, rogue peers can attempt to influence the number of optimistic unchokes by maintaining larger peer lists, which effectively increases the probability that their connection will be unchoked.

Another well-known vulnerability of the protocol pertains to free riding—in other words, selfish peers who do not contribute to the swarm. The current understanding is that BitTorrent's rate-based incentive mechanisms are relatively robust against rogue peers; however, the mechanisms can also promote free riding because there is no built-in system for rewarding good peers and punishing rogue free-riding peers. Indeed, many research proposals have focused on introducing reputation management to BitTorrent or similar protocols.

Another concern is that BitTorrent's incentive mechanism is based on the upload rates of peers (the download rates observed by the current peer), which can result in unfairness pertaining to the number of blocks traded by the peers. This opens up the possibility for rogue peers to obtain more bandwidth for themselves. Fairness scores have been proposed as a potential solution. These scores can be used to first detect unfairness and then to compensate to ensure fairness.

The BitTyrant is a BitTorrent client designed with fairness in mind. This protocol rewards those users whose upload allocations are fair [52]. The peer selection has been modified to rank all peers by their receive/sent ratios, and the selection refers those peers with high ratios.

BitTorrent's popularity has resulted in some ISPs considering this type of traffic harmful and has led them to throttle BitTorrent flows in order to free more network capacity for other applications. This development has resulted in a number of technologies for both detecting BitTorrent and P2P flows and obscuring them from being detected. Solutions such as the *protocol header encrypt (PHE)* and *message stream encryption/protocol encryption (MSE/PE)* are features of some BitTorrent implementations that masquerade the protocol traffic to make it more difficult to detect.

4.7.8 Service Capacity

Estimating the service capacity of a P2P network is important for practical deployments. Service capacity pertains to the determination of the number of initial seeds and peers and network configuration necessary for the required performance level. The aim is to avoid underprovisioning and overprovisioning the system. Flash crowds are one of the key challenges for estimating service capacity. The system should be able to handle the flash crowd effect, in which the service sees a dramatic spike in the arrival rate of requests.

To consider a simple example, a host has a popular file and many peers are requesting it simultaneously. In BitTorrent, this initial seed can be overwhelmed by the requests; however, when the file exchange progresses, other peers will become seeds and the level of traffic experienced by the initial seed levels off. When the number of seeds is large enough to satisfy all the requests in the system, the system enters a *steady state* [353].

We observe that the cooperative nature of many P2P systems, including BitTorrent, can significantly alleviate bandwidth demands and processing requirements and thus increase service capacity of a service provider. The so-called self scaling is a desirable feature for a distributed system.

The two interesting components of service capacity are the transient and steady state. In the former, the capacity estimation pertains to the build-up phase—for example, determination of the duration from initial conditions to a steady state. This phase includes flash crowds, in which the traffic grows rapidly. The transient state is server constrained. In the steady state, the system is demand, constrained, and we are interested in estimating the average metrics—for example, average throughput and delay.

The service capacity for the two components depends on a number of issues:

- Peer selection: The peer selection is an important part of the BitTorrent protocol (along with piece selection). Peer selection is about selecting peers who are willing to share files back to the current peer. A peer-selection mechanism may take into account various issues, such as load balancing, network topology, throughput, and fairness.

- Data management: A file may be divided into a number of parts. This facilitates concurrent downloads. The granularity of this partitioning (piece size) and distribution of the parts are crucial in determining system efficiency.

- Access and scheduling policy: The number of simultaneous downloads and uploads can be monitored and controlled. This control is part of the access and scheduling policy. This policy can be used to differentiate between peers.

- Traffic: The traffic patterns resulting from requests for files and pieces and the corresponding responses. Peer life cycle and dynamics also affect the traffic patterns.

The above issues are not independent of each other, but rather they interact in many ways. A peer-selection algorithm may prioritize peers who give back to the community by uploading pieces. This kind of altruistic behavior can then result in improved service capacity because, due to the upload behavior, there will be more seeds for the files.

Experimental results indicate that seed provisioning in BitTorrent is crucial to the choking algorithm's effectiveness. The seed should be at least as fast as the fastest downloaders in order to support a robust torrent during the startup phase [195].

4.7.9 Fluid Models for Performance Evaluation

BitTorrent performance has been analyzed in the literature using analytical models, including stochastic and fluid models, extensive simulation experiments, experiments on distributed testbeds (PlanetLab), and by obtaining traces from real clients. Both analytical and empirical evaluation and estimation are needed to dimension deployments to meet the service capacity demands. Fluid models can be used to estimate analytically the protocol performance and understand the time-evolution of the system by using differential equations.

A variety of different arrival processes for new peers have been considered in the literature. The three key scenarios are as follows:

- The steady flow scenario used above assumes that new peers appear with a constant rate [224, 263, 352].
- The flash crowd scenario considers the case where a (large) number of peers appear at the same time [238], after which no new peers arrive.
- In a third scenario, the arrival rate is high in the beginning but smoothly attenuates as time passes [154, 315].

In this section, we consider a simple fluid model devised by Qiu and Srikant to study the performance of BitTorrent-like file-sharing systems [263] under a steady flow arrival scenario (the first scenario). This model consists of equations that correlate the average number of seeds, the average number of downloaders, and the average downloading time with the downloader arrival/leaving rate, the seed-leaving rate, and the per-node uploading bandwidth. Although the model includes a number of key parameters, it does not take into account such issues as the number of peering neighbors and the seeds uploading capability.

The Qiu and Srikant model describes the time-evolution of the system by differential equations. The work complements earlier studies, especially the Markovian model by Yang and de Veciana [352]. While Yang and de Veciana only use their model for numerical studies, the fluid model can be used for stability and steady-state analysis. Yang and de Veciana implicitly assume that the system is upload-constrained (i.e., $c \gg \mu$); however, Qiu and Srikant allow any positive values of c and μ. They also have an additional parameter θ modeling the rate at which downloaders abort the file transfer.

The key parameters of the models are the arrival rate of new peers, λ, the efficiency of P2P file sharing, η, and the departure rate of seeds, γ, which is a measure of selfishness of the peers. The efficiency parameter η combines the effect of the piece selection policy, the number of downloading connections, and the number of pieces.

Qiu and Srikant conclude that $\eta \approx 1$ whenever the number of pieces is sufficiently high. Both models also assume a homogeneous peer population with joint download and upload rates, c and μ, respectively.

Assuming that $\theta = 0$, the fluid model by Qiu and Srikant [263] can be represented as follows:

$$x'(t) = \lambda - \min\{cx(t), \mu(\eta x(t) + y(t))\},$$

$$y'(t) = \min\{cx(t), \mu(\eta x(t) + y(t))\} - \gamma y(t). \tag{4.1}$$

The two state variables are the number of downloaders, $x(t)$, and the number of seeds, $y(t)$. The global stability of the fluid model is examined in [262] and extension of the Qiu Srikant model to heterogeneous are considered in [210, 356].

4.8   Cross-ISP BitTorrent

The BitTorrent protocol is oblivious of the underlying network topology. On one hand this simplifies the protocol design and makes all peers equal on the overlay network in terms of peer selection. On the other hand, it does not take the underlying network topology into account and, more importantly, the underlying dominant economical models for internetworking. Indeed, this unawareness of the underlay has resulted in many ISPs throttling or limiting BitTorrent traffic in their networks.

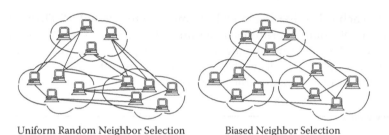

Uniform Random Neighbor Selection　　Biased Neighbor Selection

FIGURE 4.8
Biased neighbor selection.

Since BitTorrent is a very popular protocol and performs relatively well, there is a motivation to develop small changes to the protocol to make it better aware of the underlay. The question here is whether these changes affect the performance of the protocol. Recent research results indicate that, with small changes to the protocol, unnecessary interdomain traffic can be reduced significantly without affecting the performance of the protocol [29, 75].

A technique called *biased neighbor selection* has been proposed for reducing cross-ISP traffic [75]. In this mechanism, a BitTorrent peer chooses most of its neighbors from the local ISP and only a few peers from other ISPs. Essentially, the peer selection is biased toward local peers. This is illustrated in Figure 4.8.

A parameter k represents the number of external peers from other ISPs. The tracker is modified to select $35 - k$ internal peers and k external peers that are returned to the client requesting a peer list for a torrent. If there are less than $35 - k$ internal peers, the client is notified by the tracker to try again later.

The biased-neighbor selection technique works well with the rarest-first replication algorithm of BitTorrent; however, other piece-selection algorithms, such as random selection, may not lead to optimal performance.

> The tracker can use Internet topology maps or IP to autonomous system (AS) mappings to identify ISP boundaries. ISPs wishing to preserve traffic locality can also publish their IP address ranges to trackers.

Biased neighbor selection can be introduced by ISPs in a transparent fashion by using so-called *traffic-shaping devices* at the edges of their networks. These devices are located at the edge, and they can perform deep packet inspection to identify P2P traffic and possibly manipulate this traffic. Since BitTorrent is based on HTTP, it is relatively easy to detect the protocol. This means that HTTP proxies can be used as traffic-shaping devices in the BitTorrent case.

The devices could track peers inside an ISP and modify responses from the tracker when needed, in the typical case replacing outside peers with internal peers. When it is necessary to change a peer's neighbors, the device sends a TCP RESET packet on the connection between the internal and external nodes. This makes the internal peer contact the tracker for new neighbors.

An alternative biased peer-selection strategy has been proposed that relies on independent observations against well-known beacon sites. This approach does not require any explicit topology information or additional infrastructure. The key insight in this approach is that the information necessary for peer selection is already being collected by CDNs, which use dynamic DNS redirection to forward clients to low-latency servers. The assumption is that if two clients are sent to a similar set of servers, they are likely to be close to these servers and to each other. Extensive real-life measurements with BitTorrent indicate that this biased peer-selection algorithm can significantly reduce cross-ISP traffic. The results show that over 33% of the time the algorithm selects peers along paths that are within a single AS. The results also indicate that this selection results in more high-quality paths between the peers and thus better transfer rates [75].

4.9 Freenet

The unstructured P2P systems presented so far in this chapter do not offer good security and privacy features. Many of these shortcomings are addressed in the Freenet file-sharing system [82, 83].[3] This system emphasizes anonymity in file sharing and protects both authors and readers. The system works in a somewhat different way than Gnutella because it allows users to publish content to the P2P networks and then disconnect from the network. The published content will remain in the network and be accessible for users until it is eventually removed if there is not enough interest in the data. The Freenet network is responsible for keeping the data available and distributing data in a secure and anonymous way.

The developers of the Freenet network argue that true anonymity is necessary for freedom of speech. The argument is that the beneficial uses of the technology outweigh the possible negative uses. The central aim is to remove the possibility of censorship on any data. To this end, the system is built in such a way that there are no central servers and the system is not administrated by any single individual or organization. Data is stored in an encrypted format and replicated across the Freenet nodes. In order to achieve anonymity, a file is split into pieces and the pieces are then encrypted and distributed. Therefore it is very difficult for a single Freenet node to inspect the contents it is hosting. It is also very difficult to determine which hosts are providing certain files on the network.

4.9.1 Overview

The Freenet network is a decentralized loosely structured overlay network similar to Gnutella. The system is a self-organizing P2P network and creates a collaborative virtual file system by pooling unused disk space. Prominent features of the system include emphasis on security, publisher anonymity, and deniability. Moreover, the system also focuses on data replication for availability and performance.

In order to implement a distributed secure file storage and exchange service, each node in the network maintains a local data store. This store is made available to the network for reading and writing. In addition, each node maintains a dynamic routing table to be able to process requests for certain files. In order to obtain a file, a user sends a request message that includes a key for the desired file. This request also includes a timeout value, called *hops-to-live*, that is very similar to the TTL used in the Gnutella protocol. The motivation for the timeout is to detect and prevent routing loops.

The Freenet network consists of three crucial parts:

- Bootstrapping, which pertains to how a new node enters the network.
- File identifier keys, which are needed to be able to find files in the network. The keys can be derived using three different ways and each of them have their implications for the system and security.
- Key-based routing, which is the process of finding a node that hosts the desired file.

---

[3] http://freenetproject.org/

Freenet uses a variation of the MIX-net scheme that requires messages to travel through node-to-node chains. Each link in the chain is individually encrypted. Each node knows only about its immediate neighbors.

> MIX routes and forwards messages from several senders to several receivers in such a way that no relation between any particular sender and any particular receiver can be discerned by an external observer [69]. The classic application of MIX has been untraceable digital pseudonyms. Other application cases are synchronous and asynchronous communication systems, as well as electronic voting systems. Most applications use a cascade of MIXes, forming so-called *MIX-net*. MIX-nets obfuscate the relation between the senders and receivers. Each message typically goes through several MIX-stages before the ultimate destination. Onion routing (for example, Tor, presented in Chapter 9) is based on this idea.

Freenet is built around three different types of information: keys, addresses of other Freenet nodes, and the data corresponding to those keys. Freenet uses unique binary keys to identify files. If a node receives a request for a key it has, it can send a response and the file data back to the requestor. Upon receiving a request for an unknown key, a Freenet node forwards the request to another node that has keys closer to the requested key. Results for successful and failed requests are backtracked on the reverse path of the request message. When a file cannot be located, the so-called upstream nodes can then try alternative downstream nodes in order to locate the content. This routing behavior has been called *steepest-ascent hill climbing search with backtracking*.

Freenet has the following central messages:

- Data insert: This message allows a node to insert new data into the network. The message includes a key and the data file.
- Data request: A node request for a certain file. The request contains the key of the file.
- Reply: The reply is sent by the node that has the requested file. The actual file is included in the reply message.
- Data failed: This operation denotes a failure to locate a file. The message will contain the location of the node where the failure occurs and the reason.

As a result of the backtracking and trying to find alternative paths to the file, the routing accuracy and performance may improve over time. The algorithm results in the clustering of similar keys to the same nodes. The backtracking part of the algorithm allows the nodes to become better informed on when the keys nodes are hosting. Moreover, as a result of a successful file request, the requesting node will have a copy of the file. The expectation is that a node will most likely download files with similar keys, thus contributing to the scalability of the system.

Figure 4.9 illustrates a typical request sequence in Freenet. In the first step, the node *A* sends a request for a certain file. The node needs to know the key of the file. There are three basic key types, which are based on the idea of hashed identifiers. In all of the cases, the node *A* needs to obtain some information about the file, such as a short descriptive text or a public key, in order to be able to create the request.

In the following steps the request is propagated in the network, from first *B* to *C*. Node *C* does not know any other nodes and responds to *B* that the request has failed. Only after this failure *B* contacts *E*. *E* receives the request and forwards it to *F*. At this point, *F* does not know any other nodes and responds with a failure, which prompts *E* to contact *D*,

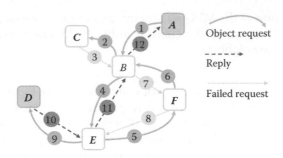

FIGURE 4.9
Overview of Freenet.

which has the requested file. The file is then sent back to *A* via *E* and *B*. *A* can then decrypt the file. At each step the validity of the file is checked by intermediate nodes, and they may cache the file.

4.9.2 Bootstrapping

Each Freenet node is connected to a set of neighbor nodes and is able to send and receive messages with them. In a similar manner to the Gnutella protocol, the way nodes discover the entry point into the network is not part of the protocol. Typically, this is done using out-of-band lists of peers.

The 0.7 version of the protocol supports two operating modes, the opennet and darknet. The former allows connections between arbitrary Freenet nodes, whereas the latter limits communications to known peers. Darknet is therefore more secure against attacks but requires laborious manual configuration. The darknet mode is built on the assumption that the user can identify trusted peers.

After a node has a peer address, it can build a secure channel over TCP/IP to that peer and start the Freenet handshake. The 0.7 version of the protocol switched to UDP to make the protocol more compatible with firewalls, as UDP allows easy hole punching. Upon receiving a handshake message, if the peer is active it can send a handshake reply back to confirm that it accepted the connection request. A connection is then active between the peers, which typically lasts for a few hours.

A node new in the Freenet network needs to have a public-private key pair and obtain a location identifier for itself. The node location is a number between 0 and 1, and it is derived through an announcement process. The process starts when a new node is announced in the network. The announcement message contains the public key and an address to an existing node obtained by out-of-band means. This announcement message is propagated by Freenet nodes. The message is propagated by randomly selecting a destination in the current node's routing table. The message has a TTL value, which determines when the message propagation is stopped. When the message propagation stops, the nodes in the chain collectively assign a new location ID for the new node. A cryptographic protocol for shared random number generation is used to prevent any participant from affecting the result. The procedure assigns the new node some subspace of the keyspace.

4.9.3 Identifier keys

The Freenet uses semantic-free references to make the keys independent of the properties of the files. This is achieved by using hash-based keys. Thus the key namespace is flat and devoid of semantics. Hashing also ensures uniform distribution of keys over the namespace.

Freenet supports three basic types of keys, the simplest of which is based on applying a hash function on a descriptive text string that is included with each file stored in the network by its creator. The *content hash key (CHK)* is the most important of the keys, and all files over 1 KB in size are divided into one or more 32 KB CHKs. The filename of a CHK is determined by its contents; it is therefore data-centric. The *signed subspace key (SSK)* is the other important type of key. SSKs feature both a public key and a human-readable filename. Essentially they are a form of self-certifying labels. The SSKs are used to represent Freenet Web sites, called *freesite*.

All the mentioned key types use symmetric encryption for the file and separate the encryption key from the actual data. The motivation for using encryption keys is that a Freenet user can deny any knowledge of having the file in their cache. The users do not know the file descriptions and decryption keys and thus cannot inspect the file contents.

A CHK is a SHA-256 hash of an encrypted resource, which makes it easy for a client to check that a downloaded resource is correct by simply hashing it, thus computing a digest, and then comparing the digest with the CHK. The CHK key is unique and provides resistance to data tampering. If a malicious node alters data, the CHK key of the data will change. As an important observation, the same data will result in the same CHK.

The SSK keys are based on public-key cryptography. The DSA algorithm is used to generate key pairs for data publishers and sign resources associated with the keys. The signature and the data can then be verified by downloaders. This means that they can authenticate the data. SSKs support a pseudonymous identity on the Freenet that can be authenticated. SSKs have been partly superseded by the *updatable subspace keys (USKs)*, which extend SSKs to allow for links that point to the most up-to-date version of a site.

The following steps illustrate how SSKs are used in Freenet:

1. The publisher generates a cryptographic keypair: a private key for signing files and a public key for verifying the signature.
2. The publisher generates a single symmetric key that is used to encrypt and subsequently decrypt the file.
3. The file is encrypted with the symmetric key and signed with the private key. The resulting signature, including the public key signature, is stored with the file, and the file is then published on Freenet. We note that neither the symmetric nor the private key are included in the file. The Freenet nodes will not have any knowledge regarding the contents of the file; however, they can verify the data matches with the signed digest.
4. The SSK consists of a hash of the public key and the symmetric key. The hash of the public key is used to locate the data on the Freenet network. The symmetric key is used to decrypt the file once it has been downloaded.

Keyword signed keys (KSK) is a variant of the SKS and is the most basic and insecure method of generating keys in Freenet. It simply consists of a human readable name (for example, KSK@test.com). In this approach, the key pair is generated in a standard way from the descriptive human-readable string. The downloader needs to know the string in order to be able to find the data and subsequently decrypt it. The approach is more lightweight than the KSK; however, it is also less secure and does not support a pseudonymous identity system. The limitation of these keys is that they are subject to spamming.

The following steps illustrate how KSKs are used in Freenet:

1. A deterministic algorithm is used to generate a cryptographic public/private key-pair and a symmetric key based on the file description. The same description will result in the same keys irrespective of the node performing the computation.
2. The public key is stored with the data and will be used to verify the authenticity of the data.
3. The file is encrypted using the symmetric encryption key.
4. The private key is used to sign the file.
5. In order to retrieve the file, a user needs to know the file description. This description can then be used to generate the decryption key.

The following table outlines the different applications of the above Freenet key types:

- CHKs are useful for single nonmutable files—for example, audio and video files.
- SSKs are intended for sites with mutable data. A typical usage case involves a Web site that has components that change over time—for example, a news section. In this case it is important to be able to authenticate the contents of the site so that malicious entities cannot change the contents. This is ensured using public-key cryptography.
- USKs are used for creating a link to the most current version of an SSK site. They are essentially wrappers around SSKs.
- KSKs are used for human-understandable links that do not require trust in the creator. These keys are vulnerable to spamming.

The SSK defines a personal namespace that anyone can read but only its owner can write to the space. Adding or modifying a file in SSK-defined namespace therefore requires the private key in order to generate a signature. SSKs define a pseudonym-based identity system. This approach can be used to send out newsletters, Web sites, and also for e-mail.

4.9.4 Key-based Routing

The system utilizes a key-based routing protocol that bears semblance to the distributed hash tables that are examined in Chapter 5. In key-based routing algorithms, files (and possibly other data) are identified using, typically probabilistically, unique keys. The routing tables that peers maintain are built in such a way that nodes can forward queries based on the keys their neighbors advertise.

There are significant differences between Freenet protocol versions. Before version 0.7, the system used a heuristic algorithm where nodes did not have fixed locations and routing was based on finding the closest node that advertised a given key. Upon successful request, new shortcut connections were sometimes created between the requesting node and the responder, and old connections were discarded.

When the 0.7 version was developed, Oskar Sandberg's research on routing in small-world networks indicated that the process of creating shortcuts, called *path folding*, is critical for scalability and efficiency. The insight was that a very simple routing algorithm could be sufficient if it uses path folding [283]. The limitation of the path folding technique in which nodes opportunistically try to find new connections is that an attacker can find Freenet nodes and connect to them. This is addressed in the Freenet version 0.7, which supports the two modes, namely Opennet and Darknet.

The new algorithm introduced the notion of node location, which is a number between 0 and 1. This location metric is used to cluster nodes. The system works as follows. When a client issues a request for a file, the node first checks if the file is locally available in the data store. If the file is not found, the file key is turned into a number in a similar fashion. The request is then routed to the node that has the numerically closest location value to the key. This routing process is repeated until a preset number of hops is reached. If the desired file is found during the routing process, the file is cached on each node along the path (given that there is room). This kind of approach works well with popular data; the more a file is requested by clients, the more it will be cached by intermediate nodes.

The above process is also used to insert a document into the Freenet network. The data is routed according to the key until the hop count is exceeded, and, if an existing document with the same key is not found, the new data is optionally stored on each node. It may happen that there is an older version of the data already in the network; if this is detected, the older data is returned to the node that sent the insertion message and it is said that the *insert collides*.

The Freenet routing algorithm relies on the properties of a small-world network for efficiency and scalability. Given that the network has the small-world properties and that the nodes constantly attempt to minimize their distance to their neighbors using path folding (also called location swapping), the Freenet network should find data in an efficient manner. The expectation is that with these assumptions data would be found on the order of $O(\log(n)^2)$ hops. Recent results indicate that this requires routing table sizes of $\Theta(\log(n)^2)$ entries [359]. In practice, the network may not be as efficient, and it does not guarantee that the data will be found.

The routing and location algorithm results in four key properties:

- Over time nodes tend to specialize in requesting for similar keys as they receive search requests from other nodes for similar keys.

- As the consequence of the above, nodes tend to store similar keys over time. This stems from the caching of requested files.

- Keys are semantic-free and the similarity of keys does not result in similarity of the files.

- Higher-level routing is independent of the underlying network topology.

4.9.5 Indirect Files

In a typical usage, the SSK keys are used in an indirect fashion by storing files that contain pointers to CHKs instead of the actual data content. These files are called *indirect files*, and they combine the human readability and publisher authentication aspects of SSKs with the efficient verification of CHKs. Indirect files support mutable data while preserving the integrity of the references.

In order to update an indirect file, the publisher first creates a new version of the file. This file will receive a new CHK due to the changed contents. The publisher can then update the SSK to point to this new version. This new file therefore replaces the old in the SSK. This technique works also for one-to-many references, and thus it is useful in splitting large files into multiple pieces.

Figure 4.10 illustrates the use of indirect files. The diagram illustrates a regular file with key "B622A17E28." The publisher first inserted the file and then a set of indirect files. The indirect files are named after the search keywords for the actual data. The indirect files are distributed across the network and do not contain the actual data, but rather a reference pointing to the data.

4.9.6 API

The Freenet system has been designed with modularity in mind. The core application is responsible for connecting to the network and acting as a proxy. The proxy provides an

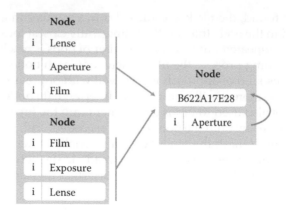

FIGURE 4.10
Indirect files in Freenet.

open application interface (API) to applications.[4] The API can be used by applications to implement different kinds of services, such as Web sites, file sharing, instant messaging, and message boards. The API implementation is text-based and it is interfaced with a TCP connection, typically by local applications.

The API supports the following key functions:

- Data insertion to the network.
- Data retrieval from the network.
- Querying the status of the network.
- Managing the other Freenet nodes that are connected to the local node.

4.9.7   Security

As mentioned previously, meeting various security challenges and concerns has been the motivation for the Freenet system. The system emphasizes anonymity and privacy of its users. Privacy is realized using a variation of Chaums mix-net scheme for anonymous communication. Messages travel through the network through node-to-node chains. Each link is individually encrypted. Each node in this chain knows only about its immediate neighbors; the endpoints are decoupled from each other. This approach protects both the publishers and the consumers. It is very difficult for an adversary to destroy a file because it is distributed across the network.

The current Freenet implementation uses a number of security solutions to improve security—for example, an outer symmetric encryption layer in communications and 256-bit Rijndael for symmetric encryption. There are also some active open security issues, such as vulnerability to file request correlation attacks and the security of the swapping algorithm. The implemented location swapping is not secure and an attacker can attempt to damage the network by using bogus swap requests. In addition, a large part of the network topology is exposed through location swapping. The swap requests are routed randomly for 6 hops, and thus intermediate nodes can see information pertaining to the source and responder.

---

[4] http://wiki.freenetproject.org/FreenetFCPSpec2Point0

4.10 Comparison

This chapter presented a number of well-known unstructured P2P algorithms that addressed decentralized and efficient data sharing over the Internet. Unstructured P2P algorithms can be seen as part of the first-generation P2P overlay systems, and the structured systems examined in Chapter 5 are examples of second-generation algorithms.

We observed that in many cases it is beneficial to incorporate some structure in a basic unstructured P2P overlay. For example, the later versions of the Gnutella protocol utilize ultra nodes that better leverage the power-law nature of the network. This ultra node variant of the basic Gnutella algorithm is expected to perform much better in terms of networking costs and hops needed to find resources. In a similar fashion, the original Freenet flooding-based routing algorithm was extended to support path folding, which in essence creates shortcuts across the network [283].

These newer versions of Gnutella and Freenet can be said to be hybrid P2P systems in the sense that they combine features from structured systems and utilize a loosely built structure (for example, ultra nodes and path folding) in order to make the network more efficient and scalable [11, 213]. These systems are not fully structured, because they do not enforce strict rules for the placement of keys and data on nodes. Fully structured systems can ensure that a data item can be found in bounded steps, typically logarithmic to the number of nodes. Structured and hybrid systems do not in general provide such guarantees. Hybrid systems based on high-capacity superpeers can be used to provide many of the advantages of a centralized system while still retaining good scalability [242].

Figure 4.11 presents a comparison of the unstructured P2P algorithms presented in this chapter. The algorithms are compared based on their properties—namely, decentralization, foundation for the distributed operation, routing function, routing performance, routing state, and reliability. In terms of decentralization, the discovery part of BitTorrent is centralized with the tracker coordinating the file exchange process. This system does not maintain multihop routing tables as such, but instead relies on the tracker for exchanging peer information. Since this protocol is centralized, data (identified by the torrent file) can be located. BitTorrent has several mechanisms for ensuring reasonable fairness in the system. The two key mechanisms are the peer-selection and the piece-selection algorithms. BitTorrent is reliable given that the initial seed and tracker are available. The two main challenges for BitTorrent are free riding (fairness) and taking ISP considerations into account.

The two other key unstructured protocols discussed in this chapter were the Gnutella and Freenet protocols. Interestingly, the early versions of these protocols were unstructured in the strict sense, and the later versions have been modified to incorporate some structure. Therefore, the later versions do not suffer from the scalability problems introduced by the flooding mechanism in the same way the earlier versions did. Both Gnutella and Freenet do not guarantee that a file is found, and in this way they differ from BitTorrent, which, due to its centralized nature, can guarantee that all pieces can be found.

A number of extensions and modifications have been proposed in the literature for making unstructured P2P systems more scalable. Example solutions include replacing flooding with random walks and adding replication support. More sophisticated broadcast policies can also be used to enhance system performance. Yang and Garcia-Molina have proposed using broadcasting policies that rely on past history to select the neighbors to which a query should be forwarded. Local indices are another technique that can be used, which maintains an index of the data stored by nodes within some radius from the current node [351]. The index can then be used to forward data to nodes that are likely to be in the direction

	BitTorrent	Freenet v0.7	Gnutella v0.4	Gnutella v0.7
Decentralization	Centralized model	Similar to DHTs, two modes (darknet and opennet), two tiers	Flat topology (random graph), equal peers	Random graph with two tiers. Two kinds of nodes, regular and ultra nodes. Ultra nodes are connectivity hubs.
Foundation	Tracker	Keywords and text strings are used to identify data objects. Assumes small world structure for efficiency	Flooding mechanism	Selective flooding using the super nodes.
Routing function	Tracker	Clustering using node location and file identifier. Searches from peer to peer using text string. Path folding optimization	Flooding mechanism	Selective flooding mechanism
Routing performance	Guarantee to locate data, good performance for popular data	Search based on Hop-To-Live, no guarantee to locate data. With small world property $O(\log(n)^2)$, hops are required, where n is the number of nodes.	Search until Time-To-Live expires, no guarantee to locate data	Search until Time-To-Live expires, second tier improves efficiency, no guarantee to locate data
Routing state	Constant, choking may occur	With small world property $O(\log(n)^2)$	Constant	Constant
Reliability	Tracker keeps track of the peers and pieces	No central point of failure	Performance degrades when the number of peers grows	Performance degrades when the number of peers grows

FIGURE 4.11
Comparison of unstructured P2P algorithms.

where desired content is stored. Simulation studies indicate that the use of the local indices can improve the performance compared to flooding by one to two orders of magnitude [88, 162].

Chawathe et al. [70] have proposed the Gia system, which addresses Gnutella's scalability issues. This proposal combines a number of optimization techniques—namely, topology adaptation that identifies high-capacity nodes and maintains a short distance to them, active flow control that voids hotspots, and a search protocol based on random walks. The simulation results of the system indicate that overall system capacity can be increased by three to five orders of magnitude.

The connectivity properties and reliability of unstructured P2P networks have been an active research topic. Gnutella has been shown to be a highly robust overlay in the face of random breakdowns. For a realistic maximum node degree of 20, the overlay is partitioned into fragments when more than 60% of the nodes fail. Here the nature of the node failures is of course a decisive factor—are they random or part of a coordinated attack against the overlay [284]? Gnutella exhibits the properties of a power-law network, in which most nodes have few links and only a few nodes are highly connected. This means that many of the less connected nodes can be removed without affecting connectivity of the network graph. On the other hand, if a highly connected node is removed, the network may become partitioned. Indeed, power-law networks are vulnerable to attacks against the busy hubs, but they are robust against random node attacks.

Putting the above observations together, we can summarize that unstructured P2P networks have favorable properties for a class of applications. The applications need to be willing to accept best-effort content discovery and exchange and to host replicated content and then share the content with other peers. The peers may come and go, and the system state is transient (minimal assumptions on how long each peer participates in the network). The dominant operation in this class of applications is keyword-based searching for content.

5

Foundations of Structured Overlays

This chapter examines the foundations of structured overlay technologies that place more assumptions on the way nodes are organized in the distributed environment. We examine different geometries for the basis of distributed hash tables (presented in the next chapter) and analyze early solutions such as consistent hashing and linear hashing for distributed files (LH*).

5.1 Overview

The two usage environments for overlays are clusters and wide-area environments. These two environments are radically different. Clusters can be assumed to be under a single administrative domain, which is secure, predictable, and engineered to avoid network partitions. Clusters have low-latency and high-throughput connections.

Wide-area environments such as the Internet are unreliable and subject to connectivity problems, bandwidth and delay limitations, and network partitions. Moreover, there are multiple administrative domains, and the operating environment is inherently insecure.

In both clusters and wide-area environments, structured overlays characteristically emphasize the following properties:

- Decentralization. The nodes collectively form the system without central coordination.
- Scalability. The system should perform efficiently, even with millions of nodes.
- Fault tolerance. The system should be reliable, even with nodes continuously joining, leaving, and failing.

Structured overlays are typically based on the notion of a semantic-free index [88, 162, 339]. They utilize hashing extensively to map data to servers. This mapping can be done directly to a set of servers, as in the case of the cluster-based techniques used in LH* and Ninja, or the mapping can be done hop-by-hop by comparing addresses derived using hashing or other randomization techniques. The cluster-based techniques typically can guarantee a very small number of hops to reach a given destination. The decentralized DHTs discussed in the next chapter, on the other hand, balance hop count with the size of the routing tables, network diameter, and the ability to cope with changes.

Characteristically, hashing-based techniques such as the ones discussed in this and the next chapter do not understand the semantics of the data [156, 275]. This is an obvious limitation of structured approaches compared to unstructured ones. In unstructured systems, the peers can support complex query processing, which is more difficult with semantic-free indexing.

With semantic-free indexing in structured overlays, data objects are given unique identifiers called keys that are chosen from the same identifier space. Keys are mapped by the overlay network protocol to a node in the overlay network. The overlay network needs to then support scalable storage and retrieval (key, value) pairs.

In order to realize the insertion, lookup, and removal of (key, value) pairs, each peer maintains a routing table that consists of its neighboring peers (their node identifiers and IP addresses). Lookup queries are then routed across the overlay network using the information contained in the routing tables. Typically, each routing step takes the query or message closer to the destination.

We distinguish between a *routing algorithm* and the *routing geometry*. The *algorithm* pertains to the exact details of routing table construction and message forwarding. *Geometry* pertains to the way in which neighbors and routes are chosen. Geometry is the foundation for routing algorithms [152]. We start this chapter by examining various geometries for structured overlay networks. The key observation is that the geometry plays a fundamental part in the construction of decentralized overlays.

After examining the geometries, we consider some of the fundamentals of overlays and DHTs, namely consistent hashing and cluster-based approaches to distributed data structures.

Consistent hashing was first introduced in 1997 as a solution for distributing requests to a dynamic set of Web servers [179]. In this solution, incoming messages with keys were mapped to Web servers that can handle the request. Consistent hashing has had dramatic impact on overlay algorithms. Indeed, most algorithms presented in this and the following chapter are based on this technique. DHTs utilize consistent hashing to partition an identifier space over a distributed set of nodes. The key goal is to keep the number of elements that need to be moved at minimum.

The Ninja system was designed to support robust distributed Internet services. One key component of the system was a cluster of servers for scalable service construction [26, 149]. A *distributed data structure (DDS)* is a self-managing storage layer that runs on a cluster. The aim of the DDS is to support high throughput, high concurrency, availability, and incremental scalability, and to offer strict consistency guarantees for the data.

The concept of a *scalable distributed data structure (SDDS)* was presented in 1993 [206, 208]. It was designed for cluster-based scalable file access. Part of the SDDS nodes are clients and part are servers storing data in buckets and addressed only by the clients.

5.2 Geometries

Parallel interconnection networks have been the subject of extensive research over the past decades. This work has resulted in a collection of topologies over static graphs. The main application area of the interconnection networks has been the development of efficient hardware systems, typically switches. Many of these interconnection geometries are now being considered for overlay networks as well. The new challenges include dynamic and random geometries [219].

The five frequently used overlay topologies are *trees*, *tori* (*k*-ary *n*-cubes), *butterflies* (*k*-ary *n*-flies), *de Bruijn graphs*, *rings*, and the *XOR geometry*. In this section, we briefly outline these topologies. They are revisited when presenting the DHT algorithms in the next chapter. We refer to some of the DHT algorithms here and provide sufficient detail to highlight how the geometries relate to the structural properties of the DHTs. The differences between some of the geometries are subtle. For example, it can be seen that the static DHT topology emulated by the DHT algorithms of pastry and tapestry are Plaxton trees; however, the dynamic algorithms can be seen as approximation of hypercubes.

The two important characteristics for the geometries are the network degree and network diameter. High network degree implies that joining, departing, and failing may affect more nodes. The geometries can be grouped based on the network degree into two types: constant-degree geometries and nonconstant degree geometries. In the former, the average degree is constant irrespective of the size of the network. The latter type involves average node degrees that grow typically logarithmically with the network size. The geometries in this section can be classified into these two types. Trees, hypercubes, rings, and the XOR geometry are examples of nonconstant degree geometries. Tori, butterflies, and de Bruijn graphs are examples of constant degree geometries [264].

5.2.1 Trees

The tree's hierarchical organization makes it a suitable choice for efficient routing. One of the first DHT algorithms, the Plaxton's algorithm, is based on this geometry [257]. In a tree geometry, node identifiers represent the leaf nodes in a binary tree of depth $\log n$. The distance between any two nodes is the height of their smallest common subtree.

For scalable networking, each node maintains a routing table with $\log n$ neighbors. In this table, the ith neighbor is at distance i from the current node. Greedy routing can then be used to forward a message to its destination on the network, given the target identifier. More specifically, the routing table is constructed in such a way that a node's neighbors have a common identifier prefix with the node, and this prefix differs in the next bit. The table can then be used to forward messages toward their destinations by fixing the highest order bit on which the current node differs from the destination.

5.2.2 Hypercubes and Tori

The d-dimensional 2-ary hypercube $H(d)$ is an undirected graph $G = (V, E)$ with node set $V = [2]^d$ and edge set E that contains all edges $\{u, v\}$, satisfying the property that u and v differ in by exactly 1 bit. Figure 5.1 illustrates an example hypercube.

In general, an arbitrary k-ary n-cube can be constructed by adding dimensions in an iterative fashion. A hypercube of dimension 0 is a single point. A k-ary 1-cube is a k-node ring. Connecting k of these 1-cubes in a cycle adds a dimension, forming a k-ary 2-cube. In general, a hypercube of dimension k is constructed by connecting the corresponding vertices of hypercubes of dimension k-1. A hypercube corresponds to a collapsed butterfly where each column maps to a single vertex with $2 \log(n)$ incident edges.

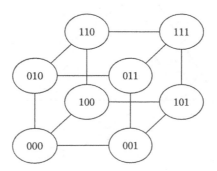

FIGURE 5.1
A hypercube.

The distance between two nodes in the hypercube geometry is the number of bits by which their identifier differs. Hypercube-based routing works in a similar manner to the above tree-geometry-based routing. At each step a greedy forwarding mechanism corrects (or fixes) 1 bit to reduce the distance between the current message address and the destination. The main difference between hypercube routing and tree routing is that the former allows bits to be fixed in any order, whereas the latter requires that the bits are corrected in a strict order.

Hypercubes are related to tori. In one dimension a line bends into a circle (a ring), resulting in a 1-torus. In two dimensions, a rectangle wraps into the two-dimensional torus, 2-torus. In a similar fashion, an n dimensional hypercube can be transformed into an n-torus by connecting the opposite faces together. The *content addressable network (CAN)* presented in the next chapter is an example of a DHT based on a d-dimensional torus.

5.2.3 Butterflies

A k-ary n-fly network consists of k^n source nodes, n stages of k^{n-1} switches, and k^n destination nodes. The network is unidirectional and the degree of each switching node is $2k$. The diameter of the network is logarithmic to the number of source nodes. At each level l, a switching node is connected to the identically numbered element at level $l + 1$ and to a switching node whose number differs from the current node only at the lth most significant bit.

The butterfly contains a binary tree with root in the first level of the butterfly network, and leaves are the nodes in the last level. Figure 5.2 presents a butterfly network with eight source nodes and highlights the binary tree rooted at one of the source nodes.

The main drawback of this structure is that there is only one path from a source to a destination; in other words, there is no path diversity. In addition, butterfly networks do not have as good locality properties as tori. A *wrapped butterfly* of dimension k can be obtained from a butterfly of dimension k by merging the first and last levels.

5.2.4 de Bruijn graph

An n-dimensional de Bruijn graph of k symbols is a directed graph representing overlaps between sequences of symbols. It has k^n vertices that represent all possible sequences of

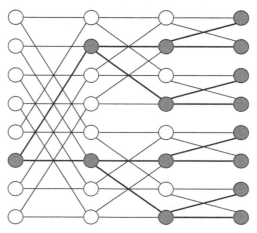

FIGURE 5.2
Examples of a butterfly network with a binary tree highlighted.

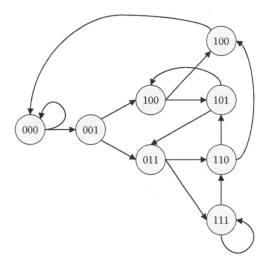

FIGURE 5.3
Examples of a de Bruijn graph (d = 3).

length n of the given symbols. The resulting directed graph has a fixed out-degree of the size of the alphabet, k, and diameter defined by n. In an n-dimensional de Bruijn graph with two symbols, there are 2^n nodes, each of which has a unique n-bit identifier (Figure 5.3). The node with identifier i is connected to nodes $2i \bmod 2^n$ and $2i + 1 \bmod 2^n$. A routing algorithm can route to any destination in n hops by successively shifting in the bits of the destination identifier.

Routing a message from node m to node k is accomplished by taking the number m and shifting in the bits of k one at a time until the number has been replaced by k. Each shift corresponds to a routing hop to the next intermediate address. The hop is valid because each node's neighbors are the two possible outcomes of shifting a 0 or 1 onto its own address. This geometry has been used in several DHTs (for example, Koorde, presented in the next chapter, and D2B [134]).

5.2.5 Rings

Rings are a popular geometry for DHTs due to their simplicity. In a ring geometry, nodes are placed on a one-dimensional cyclic identifier space. The distance from an identifier A to B is defined as the clockwise numeric distance from A to B on the circle. Rings are related with tori and hypercubes, and the 1-dimensional torus is a ring. Moreover, a k-ary 1-cube is a k-node ring. The chord DHT is a classic example of an overlay based on this geometry. Figure 5.4 presents an example a ring geometry. Clockwise direction is highlighted. Each node has a predecessor and a successor on the ring, as well as an additional routing table for pointers to increasingly far away nodes on the ring.

Efficient routing on a ring is based on a cyclic identifier space of $2^n - 1$ identifiers. If each node maintains $\log n$ neighbors, routing on a ring can be achieved in $O(\log n)$ hops irrespective of how a node chooses its ith neighbor from the range $[(a + 2^i), (a + 2^{i+1})]$. Chord selects the exact node closest to $a + 2^i$ on the circle; however, other selection strategies result also in similar performance.

When a message is routed on a ring, the next node after the first hop has approximately $(\log n) - 1$ possible next hops. This results in a total of approximately $(\log n)!$ possible routes for a typical path.

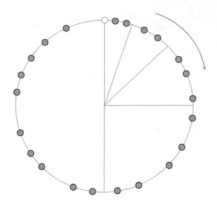

FIGURE 5.4
Example of an identifier ring.

5.2.6 XOR Geometry

The Kademlia P2P system defines a routing metric in which the distance between two nodes is the numeric value of the exclusive OR (XOR) of their identifiers [226]. We call this geometry based on the XOR metric the XOR geometry.

Kademlia routing tables are built by each node picking $\log n$ neighbors, where the ith neighbor is any node within an XOR distance of $[2^i, 2^{i+1}]$. The resulting routing tables correspond to the ones generated by the Plaxton's algorithm when there are no failures. With failures, the XOR geometry is more flexible and allows a node to choose which high-order bit to fix in order to make progress toward the destination.

More specifically, assuming two 160-bit identifiers, x and y, Kademlia defines the distance between them as their bitwise exclusive or (XOR, denoted by $\oplus$) taken as an integer,

$$d(x, y) = x \oplus y. \tag{5.1}$$

It follows that $d(x, x) = 0$, $d(x, y) > 0$ if $x \neq y$, and $\forall x, y : d(x, y) = d(y, x)$. XOR also satisfies the triangle property: $d(x, y) + d(y, z) \geq d(x, z)$. The triangle property follows from the fact that $d(x, z) = d(x, y) \oplus d(y, z)$ and $\forall a \geq 0, b \geq 0 : a + b \geq a \oplus b$.

In a similar manner to Chord's clockwise circle metric, XOR is unidirectional. For any given point x and distance $\Delta > 0$, there is exactly one point y such that $d(x, y) = \Delta$. Unidirectionality ensures that all lookups for the same key converge along the same path, regardless of the originating node. Thus, caching (key, value) pairs along the lookup path can be used to alleviate traffic hot spots. Unlike Chord, but in a similar fashion to pastry, the XOR topology is also symmetric ($d(x, y) = d(y, x)$ for all x and y).

5.2.7 Summary

Gummadi et al. compared the different geometries, including the tree, hypercube, butterfly, ring, and XOR geometries [152]. Loguinov et al. complemented this list with de Bruijn graphs. The conclusions of these comparisons include that the ring, XOR, and de Bruijn geometries are more flexible than the others and permit the choice of neighbors and alternative routes. The ring and XOR geometries were also found to be the most flexible in terms of choosing neighbors and routes. Only de Bruijn graphs allow alternate paths that are independent of each other.

A cost-based model has been proposed for comparing different routing geometries. This model focuses on estimating the resources that each node contributes to the overlay network. The key motivation for such a model is to understand the potential disincentives for nodes to collaborate in realizing the overlay. The crucial issue is how much nodes value the resources they use to forward traffic on behalf of other nodes. This model indicates that the social optimum may significantly deviate from a Nash equilibrium when nodes value their routing resources [76]. DHT topologies have also been investigated in the framework of Cayley graphs, which are one of the most important group-theoretic models for the design of parallel interconnection networks [264]. This model can provide a unifying framework for many of the geometries.

5.3 Consistent Hashing

In most traditional hash tables, a change in the number of array elements causes nearly all keys to be remapped. They are therefore useful for balancing load to a fixed collection of servers but not suitable for dynamic server collections. *Consistent hashing* [179] is a technique that provides hash table functionality in such a way that the addition or removal of an element does not significantly change the mapping of keys to elements. The technique requires only K/n keys to be remapped on average, where K is the number of keys and n is the number of nodes.

Consistent hashing was first introduced in 1997 to cope with the dynamic load with a supply of Web servers [179]. In this solution, incoming messages with keys were mapped to Web servers that can handle the request. This mapping was done using the consistent hashing technique. This allows the addition and removal of servers at runtime, with K/n elements being moved for each change. This technique can also be used to cope with partial system failures in large Web applications. As mentioned in the overview section, consistent hashing has had dramatic impact on overlay algorithms.

The technique defines a view to be a set of caches, and they are assumed to be inconsistent. Each client is aware of a constant fraction of the available views. The three important properties in consistent hashing are the smoothness, spread, and load properties:

- Smoothness: When a cache is added or removed, the expected fraction of objects that must be moved to a new cache is the minimum needed to maintain a balanced load across the caches.
- Spread: Over all the client views, the total number of different caches to which a object is assigned is small.
- Load: The total number of caches responsible for a particular object is limited.

A view is a subset of the buckets (cache servers). Consistent hashing uses a *ranged hash function* to specify an assignment of items to buckets for every possible view. A ranged hash family is said to be balanced if given a particular view, a set of elements, and a randomly chosen function from the hash family, with high probability the fraction of items mapped to each bucket is $O(1/|V|)$, where V is the view. A balanced ranged hash function distributes load evenly across the buckets.

Monotonicity is another important property for the hash function. This property says that some items can be moved to a new bucket from old buckets, but not between old buckets. Monotonicity therefore contains the essence of consistency—that elements should only be moved to preserve an even distribution. The third key property is spread, which is about ensuring that at least a constant fraction of the buckets are visible to clients.

Consistent hashing involves the construction of a ranged hash family with the desired properties. The idea is to map buckets and items to the unit interval and map a data item to the closest bucket. One point is not sufficient to characterize a bucket due to the required properties. A bucket is replicated $\kappa \log(C)$ times, where C is the number of distinct buckets and κ is a constant. When a new bucket is added, only those items are moved that are closest to one of its points.

A balanced binary search tree can be used to store the correspondence between segments of the unit interval and buckets. If there are C buckets, then there will be $O(\kappa C \log C)$ intervals and the tree will have depth $O(\log C)$. A single hash computation takes $O(\log C)$ time. The time for addition or removal of a bucket is $O(\log^2 C)$, since we insert or delete $\kappa \log(C)$ points for each bucket. The hashing time can be improved to $O(1)$ by dividing the interval into segments and keeping a separate tree for each segment [179].

Consistent hashing allows buckets to be added in any order, whereas Litwin's linear hashing (LH*) scheme requires buckets to be added one at a time in sequence. Consistent hashing has been extensively applied for DHTs, and a number of improvements have been developed—for example, for address space balancing and load balancing arbitrary item distributions across nodes [180]. Extended consistent hashing is a technique that randomizes queries over the spread of caches to reduce the load variance [196].

5.4 Distributed Data Structures for Clusters

The concept of a *scalable distributed data structure (SDDS)* was presented in 1993 [206, 208]. It was designed for cluster-based scalable file access. Part of the SDDS nodes are clients and part are servers storing data in buckets and addressed only by the clients. In this section, we first outline linear hashing, which is the basis for LH*, consider the taxonomy of SDDS structures, and then examine the LH* system in more detail. Finally, we consider the Ninja system and its approach to a cluster-based distributed data structure.

5.4.1 Linear Hashing

A hash table is a data structure that associates keys with values and supports constant time lookups on average. *Linear hashing* is a dynamic hash table algorithm proposed by Witold Litwin in 1980 [205] for extensible RAM memory or disk files. This algorithm allows for the expansion of the hash table one slot at a time. The key idea is to spread the cost of the table expansion across insertion operations. Thus the algorithm is suitable for dynamic and interactive applications.

In linear hashing, files are organized into buckets stored either in RAM or on disk. An LH file is a set of buckets. The buckets are addressed using hash functions and they are bounded by a size of a power of two. The algorithm uses a split variable i to control the expansion of the structure. The system has initially m buckets. The address function used by LH is as follows:

$$addr(l, k) \leftarrow h(k) \bmod 2^l m, \qquad (5.2)$$

where h is a hash function, k is the key, and l is the level (or split round).

Lookups using LH use this address function $addr(l,k)$ if it is greater than or equal to the split variable i. If this is not the case, $addr(l+1,k)$ is used instead. The expansion of the structure involves rehashing the entries pointed to by the split variable i to the new location indicated by $addr(l+1,k)$. The split variable therefore indicates which bucket to split next. After expansion, the split variable is incremented by one, and if it reaches 2^l, then l

is incremented by one and the split variable is reset. At any time, two hash functions are used, defined by parameter pairs (l, k) and $(l + 1, k)$. There are a total of $2^l m + i$ buckets.

5.4.2 SDDS Taxonomy

> The first SDDS was the well-known LH* algorithm that has several variants and implementations today. A notable property of LH* and some of its variants is that a key search needs at most two forwarding messages (hops) to find the correct server independent of the file size. This kind of approach lends itself well to applications that require processing of large files.

Figure 5.5 presents a taxonomy of SDDS algorithms that have been proposed in literature. Many different classifications of the structures can be presented, based on various properties. This taxonomy classifies the structures into hash-based and tree-based. Hash-based structures are further classified into one-dimensional and multi-dimensional. The LH* is the classical example of a one-dimensional SDDS. Basic LH* has been extended for high availability (LH*$_g$, LH*$_{lh}$ [181]), security (LH*$_s$ [204]), and most recently for peer-to-peer (LH*$_{P2P}^{RS}$).

LH*$_g$ [337] extends the LH* data structure with high availability by grouping records into primary files. The records in primary files are stored at different buckets (servers). A group is a logical structure of up to k records, where k is a file parameter. Every group contains a parity record allowing for the reconstruction of an unavailable element. The basic scheme

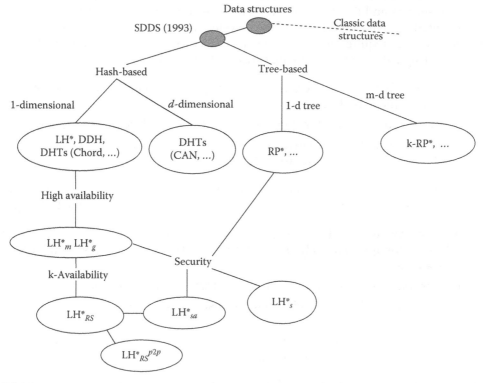

FIGURE 5.5
Taxonomy of scalable distributed data structures.

may be generalized to support the unavailability of any number of sites, at the expense of storage and messaging. LH*g file takes about $1/k$ times larger space.

LH*$_{P2P}^{RS}$ is an SDDS design intended for P2P applications. This LH* variant stores and processes data on SDDS peer nodes. Each node is both an SDDS client and potentially an SDDS server. Key-based queries require at most one forwarding message. The parity management uses a Reed Salomon erasure correction scheme to deal with churn [209].

Range partitioning (RP*) [207] preserves the key order in a similar fashion as a B-tree [90]. A RP* file is partitioned into buckets so that each bucket contains a maximum of b records, with the keys within some interval $]\lambda, \Lambda]$ called the bucket range. The parameter λ is the minimal key and Λ is the maximal key of the bucket. A record r with key c is in bucket with the range $]\lambda, \Lambda]$ only if $c \in]\lambda, \Lambda]$. A split partitions a full bucket as in a B-tree. The ranges resulting from any number of splits partition the key space and the file into a set of intervals $]\lambda_i, \Lambda_i]$, where i designates the ith file bucket. For any key c, there is only one bucket in the RP* file that may contain it.

An RP* file is designed to support range queries [207]. A range query requests all records within some range $[\lambda_R, \Lambda_R]$. The application submits a query to the SDDS client at its node. The client then receives data from the buckets that have matching items. Results are sent by each bucket i that satisfies $[\lambda_i, \Lambda_i] \cap [\lambda_R, \Lambda_R] \neq 0$.

5.4.3   LH* Overview

LH* generalizes linear hashing to decentralized distributed operation [206, 208]. The system supports constant time insertion and lookup of data objects in a cluster. Data items are hashed into buckets, with each bucket residing on a server. New servers are incorporated into the system using a split operation when a bucket overflows. A split controller manages the split operation. When a split is performed, a new server is added to the system from a supply of servers and the hashing parameters are adjusted accordingly. In a distributed environment, the clients have a view to these system parameters, which in some cases may be out of date. This requires autocorrection and synchronization mechanisms [206].

LH* was designed with the following constraints in mind:

- A file expands to new servers gracefully. Expansion is only used when existing servers are efficiently loaded.

- There is no master server that performs data (bucket) address computations.

- The file access and maintenance primitives (lookup, insertion, split, remove, ...) do not require atomic updates to multiple clients.

LH* is based on a supply of servers and supports incremental scalability for data through its linear hashing technique. The following properties are supported by the technique:

- A file can grow to a size bounded by the number of servers and their capacity, with the load factor approximately constant, typically between 65% to 95%, depending on parameters.

- An insertion usually requires one message, three in the worst case.

- A retrieval of an object given its identifier usually requires two messages, four in the worst case.

- A parallel operation on a file of M buckets costs at most $2M + 1$ messages, and between 1 and $O(\log M)$ rounds of messages.

Figure 5.6 presents an overview of the LH* system. The central components are the following: a supply of servers and m clients. Each server hosts a bucket containing the data

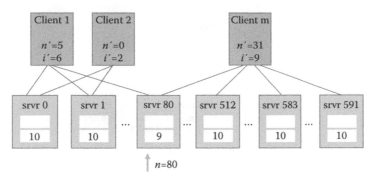

FIGURE 5.6
Overview of LH*.

items, and the clients maintain views to the global system state. n is the split pointer and i is the level. In this example, the file has expanded so that $i = 9$ and $n = 80$.

The clients are not required to have a consistent view of i and n. Every address calculation (hashing to find the correct server) starts with a client address calculation. In the figure, Client 1 believes that $i' = 6$ and $n' = 5$ and that the file has only 69 buckets. This results in a key that may not be the correct one. Thus there is a possibility for an addressing error. The LH* system solves this by also using server address calculation, in which a server can forward an incorrectly routed message. In this case the server also sends an adjustment message to the client so that the client can update its view of the two parameters.

An LH* file expands as an LH file, through the linear change of the pointer and splitting of each bucket n. The values of n and i can be maintained at a site that becomes the *split coordinator (SC)*. The splitting can be uncontrolled or it can be controlled. In the former, it is performed for all collisions and in the latter it is performed for some but not all collisions.

Server n (with bucket level l) that receives the message to split:

- Creates bucket $n + 2^l$ with level $l + 1$
- Splits bucket n applying the hash function for level $l + 1$ (roughly half of the objects are sent to bucket $n + 2^l$)
- Updates $l \leftarrow l + 1$
- Commits the split to the coordinator

Figure 5.7 illustrates the (uncontrolled) split operation. After an insert operation, the bucket c overflows and the split coordinator starts the split process. The coordinator contacts the node identifier by the split pointer n (with bucket level l). The server n creates a new bucket $n + 2^l$ with level $l + 1$, splits bucket n (part of the objects are transferred to bucket $n + 2^l$), updates $l \leftarrow j + 1$, and commits the split to the coordinator.

Splitting enables the system to maintain high utilization of buckets and servers. The limitation of the algorithm is that it requires global knowledge of the system, and that the clients know the addresses of the servers. As a reverse operation to the split, LH* also supports the merging of buckets.

The coordinator in LH* can be seen as a superpeer in P2P terminology. On the other hand, the LH* client is not a peer as such, since it is not involved in storing the file. The system can be extended to P2P environments as exemplified in the LH*$_{P2P}^{RS}$ system.

LH* can also be realized without a coordinator. In this case the insert and search costs are the same as for the basic mechanism (with a coordinator). The splitting cost decreases on the average but becomes more variable. This variability is due to cascading splits that are needed to prevent file overload.

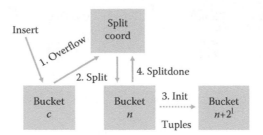

FIGURE 5.7
Splitting buckets.

5.4.4 Ninja

The Ninja system was designed to support robust distributed Internet services and to allow heterogeneous devices to access them. One key component of the system is a cluster of servers for scalable service construction [26, 149, 150]. A *distributed data structure (DDS)* is a self-managing storage layer that runs on a cluster. The aim of the DDS is to support high throughput, high concurrency, availability, and incremental scalability, and to offer strict consistency guarantees for the data. The DDS is used using a small set of API functions that abstract the cluster details from the developers. A key design goal was to make the DDS appear as a conventional data structure, such as a hash table or a tree, to the developers.

The API provides services with *put()*, *get()*, *remove()*, *destroy()* operations on hash tables. Behind the API the DDS needs to implement the mechanisms to access, partition, replicate, scale, and recover data. A distributed hash table was implemented as an example of the DDS concept in Ninja. All operations inside the distributed hash table are atomic, meaning that a given operation is either performed fully or not at all. In order to ensure reliability, elements are replicated within the DDS across multiple nodes called *bricks*. A two-phase commit algorithm is used to keep the replicas coherent. A brick consists of a buffer cache, a lock manager, a persistent chained hash table implementation, and an RPC communications facility.

Figure 5.8 presents an overview of the Ninja DDS. The bricks are located in a *storage area network (SAN)* and hosted by a number of servers. The SAN is a low-latency and high throughput network. Each brick is a single-node, durable part of the distributed hash table. The bricks are replicated across the servers. There are many clients that use the DSS API to access the data stored by the bricks.

The hash key space is split into equally sized partitions, bricks, and each partition is replicated to multiple nodes for reliability. A metadata map is used to find a key's partition. This is implemented using a trie data structure. A second map, the replica group membership map, is used to find a list of bricks that are currently active replicas in the partition's replica group. The two maps are maintained on each server in the system and updated in a lazy fashion when they are used to perform operations. The DDS library includes hashes of the maps in requests, and the bricks can detect if their maps are out of date and then request new ones.

When a read operation is issued on a key, the matching brick is found and the data is returned to the client. Due to replication, there can be many possible bricks matching the key, and any of them can be used. Writing data to a key involves updating all replicas. Consistency is guaranteed using optimistic two-phase commit.

The DDS library acts as the two-phase commit coordinator for state-changing operations. In the first phase, agreement from all replicas is obtained. In the second phase, the agreed

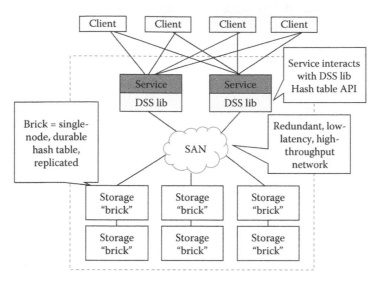

FIGURE 5.8
Overview of the Ninja DDS.

state is finalized and committed. A timeout will cause the protocol to abort. Replicas use short in-memory logs of recent-state changes for coping with interrupted updates. If a replica times out, it will contact all its peer replicas to find out if a commit has been made. If so, it will commit as well.

6

Distributed Hash Tables

In this chapter, we consider well-known distributed hash tables (DHT) solutions, such as the Plaxton's algorithm, Chord, Pastry, Tapestry, Koorde, Kademlia, CAN, Viceroy, and others. The algorithms are based on differing geometries, such as hypercubes, rings, tori, butterflies. We compare the systems and their salient features.

6.1 Overview

DHTs are based on consistent hashing examined in the previous chapter and they aim to support information lookup in decentralized Internet-wide environments. The early canonical DHT is the Plaxton's algorithm. After this the first four DHTs—namely CAN, Chord, Pastry, and Tapestry—were introduced approximately at the same time in 2001. DHTs have been an active topic both in academia and in the industry. Key examples of deployed DHT algorithms are the Kademlia used in BitTorrent, Amazon's Dynamo, the Coral Content Distribution Network, and PlanetLab. We will return to applications later in the book in Chapter 10.

In theory, DHT-based systems can guarantee that any data object can be located using $O(\log N)$ overlay hops on average, where N is the number of nodes in the system. The underlying network path between two peers can be significantly different from the path used by the DHT-based overlay network. The lookup latency in DHT-based P2P overlay networks can be high.

We briefly consider the four canonical DHTs before examining the protocols in more detail. In Chord, the participating nodes have unique identifiers and form a one-dimensional ring. Each node maintains a pointer to its successor and predecessor node (determined by their identifiers). As an optimization they maintain additional pointers (called _fingers_) to other nodes in the network. A message is forwarded greedily toward a closer node in the finger table with the highest identifier value less than or equal to the identifier of the destination node.

Pastry and Tapestry are designed after the Plaxton's algorithm and thus they are tree-based solutions. In a similar fashion to Chord, a message is routed to a node that is closer to the destination (by one more digit). Pastry uses prefix routing, whereas Tapestry is based on suffix routing.

CAN differs from the other mentioned DHTs in that it is based on a d-dimensional Cartesian coordinate space and each node has an associated zone. Each node knows its neighbors in the logical topology. Messages sent to a coordinate are delivered to the node that is responsible for the zone that contains the coordinate. Each node forwards the message to the neighbor that is closest to the destination.

An ideal DHT algorithm would meet the following requirements:

- Easy deployment over the Internet
- Scalability to millions of nodes and billions of data elements
- Availability for the data items so that faults can be tolerated
- Adaptation to changes in the network, including network partitions and churn
- Awareness of the underlying network architecture so that unnecessary communication is avoided
- Security, so that data confidentiality, authenticity, and integrity can be established and malicious nodes cannot overwhelm the overlay network

It is not easy to meet these requirements simultaneously. In this chapter, we survey the basic DHT solutions for achieving decentralized operation and scalability. Some DHT designs aim to be secure against malicious participants and to allow participants to remain anonymous. We return to security issues in Chapter 9.

6.2   APIs

The typical *application programming interface (API)* of a DHT is very simple, with only a small set of functions. We can take the cluster-based Ninja DDS API as an example of a distributed data structure API that aims to abstract the cluster details from the application software developers. A key design goal was to make the API appear as a conventional data structure, such as a hash table or a tree, to the developers. A similar goal was in LH*, which aimed to abstract the notion of a distributed file from its clients.

This same design goal is visible in later decentralized DHT systems, such as those examined in this chapter. These structured P2P systems have a common element called *key-based routing service (KBR)* [102]. This service offers efficient routing to identifiers (keys) derived from a large identifier space. The KBR abstraction is envisaged to be layered so that more complex key-based operations can be built on the basic primitives, such as the DHT API. KBR abstractions include DHTs, group anycast and multicast, and decentralized object location and routing [102].

KBR can be seen to be the common element in structured DHTs, and it implies that each overlay node maintains a routing table based on neighbors' identifiers and can route a message toward a destination identifier. As mentioned in the previous chapter, the two key parts are the routing algorithm and the geometry.

A DHT KBR is expected to provide the following operations for applications:

- join(q): current peer contacts peer q to join the system.
- leave(): current peer leaves the system.
- lookup(key): current peer searches for the peer responsible for the given key.

6.3 Plaxton's Algorithm

The Plaxton's algorithm realizes an overlay network for locating named objects and routing messages to these objects [257]. The algorithm was proposed by Plaxton, Rajaraman, and Richa in 1997 to improve Web caching performance. The algorithm guarantees a delivery time within a small factor of the optimal delivery time. The algorithm requires global knowledge and does not support additions and removals of nodes, and it is therefore a precursor to the DHT algorithms such as Chord, Pastry, and Tapestry that tolerate churn.

Plaxton's algorithm provides three operations, namely read, insert, and delete. The read operation requests an object, the insert operation creates a copy of an object that is placed in the overlay, and the delete operation removes the given object. The Plaxton overlay can be seen as a set of embedded trees in the network, one rooted in every node, where the destination is the root.

Objects and nodes have semantic-free identifiers that are independent of semantic properties. The identifiers are fixed-length sequences represented by a common base (for example, 40 hex digits = 160 bits). A hashing algorithm, such as SHA-1, is used to achieve even distribution of the identifiers in the identifier space.

> In Plaxton's algorithm, every node is assigned a unique n bit label. This n bit label is divided into l digits, with each digit having b bits, where b is the base of the addressing scheme. The routing algorithm is based on the tree geometry examined in the previous chapter and involves fixing the destination identifier from left to right digit by digit [257].

A server informs the network that it has an object available by routing a message to the root node of the object. This root node is uniquely defined for every object. The publisher sends a message to this root that is processed by all the nodes on the path from the publisher to the root. These intermediate nodes store a pointer to the server where the object can be found.

During a read operation, a client sends a message to an object. This message is initially routed through the root of the object. If the intermediate nodes have a mapping entry for the object, they redirect the read message to the server that is responsible for the object. Otherwise the message is routed to the root that is guaranteed to find an entry for the object location. Plaxton's algorithm uses a globally consistent deterministic algorithm for choosing a root node for an object.

6.3.1 Routing

The algorithm models the network as a fully connected graph with n nodes that share a set of $m = poly(n)$ objects. Every node is identified by a $\log_b n$ bitstring. In order to minimize communications cost in routing, a symmetric cost function is assumed for transferring a word from one node to another. This cost function is assumed to satisfy the triangle inequality. In order to prove correctness, the cost function was restricted to a certain family of functions that assume density bounds defined by the two constants δ, Δ. For every node, the ratio of neighbors within a given radius r and $2r$ varies at most between these constants. More formally, let M be a ball around node u with radius r, then

$$\min \delta |M(u, r)|, n \leq |M(u, 2r)| \leq \Delta |M(u, r)|. \tag{6.1}$$

Routing table of node 3642

Entries \ Levels	1 Primary neighbor	2	3	4
1	0642	X042	XX02	XXX0
2	1642	X142	XX12	XXX1
3	2642	X242	XX22	XXX2
4	3642	X342	XX32	XXX3
5	4642	X442	XX42	XXX4
6	5642	X542	XX52	XXX5
7	6642	X642	XX62	XXX6
8	7642	X742	XX72	XXX7

Wildcards are marked with X
Primary neighbor is one digit away

Example lookup

Node **3642** receives message for **2342**
• The common prefix is **XX42**
• Two shared digits, consult second column
• Send to node with one digit closer
• Fourth line with **X342**

Table size: base* address length
In this example octal base (8)
and 4-digit addresses

FIGURE 6.1
Plaxton's routing table.

Each node has a local routing table that allows it to incrementally route messages to the destination identifier digit by digit. A node N has a routing table with multiple levels, and level j represents a matching suffix up to the jth digit in the identifier. The number of entries in each level is the size of the identifier base. The ith entry at level j is the identifier and location of the closest node which ends in a shared prefix "i" with $suffix(N, j-1)$. Figure 6.1 gives an overview of the routing table for octal addressing and illustrates the message-forwarding process.

Plaxton's algorithm provides basic fault handling. Due to the suffix-matching mechanism, a message can be routed around failed nodes by choosing a similar suffix that still takes the message closer to the destination. The algorithm achieves scalability by keeping only a fraction of the routing state at each node. The digit-based routing allows for reduction of the number of candidates at each routing step by a factor of b.

6.3.2 Performance

With consistent routing tables, Plaxton's algorithm guarantees that any existing unique node in the system will be found within at most $\log_b N$ logical hops, where N is the size of the identifier namespace and b is the base. Since a node assumes that the preceding digits all match, at each level only small constant entries are maintained, resulting in a total routing table size of $b \log_b N$. It has been proven that the total network distance traveled by messages during both read and write operations is proportional to the underlying network distance.

The algorithm was designed for Web caching environments, which are typically administratively configured and involve stable hosts. As a consequence, the algorithm is not well suited for P2P environments without modifications. The Plaxton's algorithm does not provide a fully decentralized solution for peer discovery, because global knowledge of the topology is needed.

The Plaxton's algorithm has been found to have the following key limitations:

• Requirement for global knowledge
• Root nodes are possible points of failure
• Lack of ability to adapt to dynamic query patterns

6.4 Chord

The Chord protocol is a decentralized distributed hashtable algorithm for connecting the peers of a P2P network together [308, 309]. Chord is based on consistent hashing and consistently maps a key onto a node. Both keys and nodes are assigned an m-bit identifier. This identifier is a hash of the node's IP address. A key's identifier is a hash of the key.

Figure 6.2 illustrates the Chord ring. Each node has a predecessor and a successor on the ring. In addition, each node maintains the finger table to be able to reach nodes further away in the ring in clockwise direction.

> The Chord protocol is based on a logical ring with positions numbered 0 to $2^m - 1$. Key k is assigned to node successor(k), which is the node whose identifier is equal to or follows the identifier of k. If there are N nodes and K keys, then each node is responsible for roughly K/N keys.

The following theorem proved for consistent hashing [179] forms the foundation for the Chord algorithm.

THEOREM 6.1
For any set of N nodes and K keys, with high probability:

- *Each node is responsible for at most $(1 + \epsilon)K/N$ keys.*
- *When an $(N + 1)$st node joins or leaves the network, responsibility for $O(K/N)$ keys are relocated.*

The analysis for consistent hashing shows that ϵ can be reduced to an arbitrarily small constant by having each node host $\Omega(\log N)$ virtual nodes that each have their own identifier. Chord is based on these virtual nodes and assumes that each real node runs v virtual nodes. The analysis pertaining to the Chord algorithm applies to one of these virtual machines.

Chord is a popular DHT and has been used to implement a number of applications, including the following [309]:

- Cooperative mirroring: In this application, Chord is used to implement a load-balancing mechanism for distributed data across peers.
- Time-shared storage: In this application, the overlay is used for offline storage of data.
- Distributed indices: Retrieval of files over the network within a searchable database.

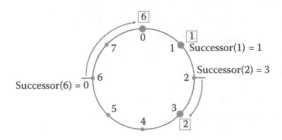

FIGURE 6.2
Overview of Chord.

- Large-scale combinatorial searches: In this application, keys are candidate solutions to a problem. The Chord mapping from the key to a node determines the node that is responsible for evaluating the candidate solution.
- Forwarding infrastructure: In this application, messages are routed and forwarded across the overlay [158, 307].

6.4.1 Joining the Network

When a new node joins the Chord network, it needs to determine its position in the Chord ring. The process starts by the new node finding a successor on the ring (based on its identifier). The three following definitions form the basic ingredients of the Chord routing [308, 309]:

- *finger[k]*: first node on circle that succeeds $n + 2^{k-1} \bmod 2^m$, $1 \leq k \leq m$
- *successor*: the next node on the identifier circle (*finger[1].node*)
- *predecessor*: the previous node on the identifier circle

Chord routing is based on two system invariants:

- Each node's successor is correctly maintained.
- For every key k, node *successor(k)* is responsible for k. A node stores the keys between its predecessor and itself. The (key, value) is stored on the successor node of key.

When a node n joins or leaves the network, responsibility for $O(K/N)$ keys takes place. To maintain a consistent mapping, when node n joins the network, some keys assigned to n's successor will be moved to n. Similarly, when n leaves the network, all keys assigned to n are reassigned to its successor.

Since the successor is known, a linear search over the network can be performed for a particular key. This linear searching is not very efficient and can be optimized by creating a routing table called the *finger table*. This table contains up to m entries. The ith entry of node n will contain the address of successor $(n + 2^i)$. Figure 6.3 presents an example of a Chord routing table.

6.4.2 Leaving the Network

A node leaving the Chord network may transfer the keys it holds to its successor before it leaves. The leaving node can also inform the successor and predecessor that it is leaving, allowing the system to reconfigure. The leaving node's predecessor will remove the node from its successor list and add the last node in the node's successor list to its own list. The current node's successor is updated in a similar fashion.

Failures and nodes leaving the system may disrupt the operation of the Chord ring. Correctness of the protocol requires that each node know its successor (invariant property). To mitigate problems with nodes leaving, Chord maintains a list of the first r successors. This successor list allows recovery when failure occurs. The probability that all r successors fail simultaneously is very small, and the system can be made more robust by increasing r. When a node leaves the system voluntarily, it will transfer its keys and notify the predecessor and successor of the leave.

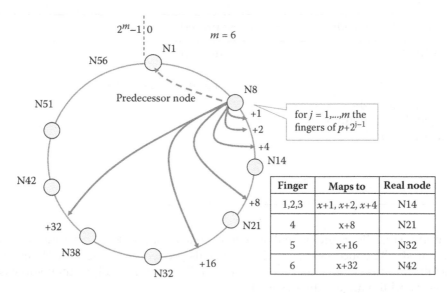

FIGURE 6.3
Example of a Chord routing table.

6.4.3 Routing

Algorithm 6.1 presents the algorithm for Chord lookup for a given key x. If x is between n and its successor, the function is finished and the successor is returned. Otherwise, n consults its finger table for the node n' that has the identifier that immediately precedes x. The *find_successor* is then invoked at n'.

Algorithm 6.2 presents the pseudocode for finding the closest preceding node using the Algorithm 6.1. This pseudocode illustrates how the finger entries are positioned at powers of two around the Chord circle. Each node can forward a query at least halfway along the remaining distance between the node and the destination. It has been proved that the number of nodes that must be contacted to find a successor in an N-node network is $O(\log N)$ [309].

Chord uses a *stabilization protocol* to ensure that the successor of each node is up to date and that the invariants are maintained. The protocol requires that when a node starts it first locate its successor by running the join algorithm. Then the node runs the stabilization protocol periodically. This protocol requests the successor s to report its predecessor p to the

---

Algorithm 6.1 Pseudocode for Chord lookup function

Data: x is the target identifier and n is the current node.
Function: *n.find_successor(x)*
/* Request node n to find the successor of x */

n.find_successor(x)
if $(x \in (n, successor)$ **then**
| **return** *successor*
else
| $n' = closest\_preceding\_node(x)$
| **return** $n'.find\_successor(x)$
end

Algorithm 6.2 Pseudocode for finding closest preceding node

Data: x is the target identifier and n is the current node.
Function: $n.closest\_preceding\_node(x)$
```
/*Search the local table for the highest predecessor of x          */
```
for $i = m$ downto 1 **do**
   | **if** $(finger[i] \in (n, x))$ **then**
   | | **return** $finger[i]$
   | **end**
end
return n

---

current node. This allows the current node to update its successor if p has just joined as the predecessor of s. The successor s can also update its predecessor if needed. When all nodes periodically run this stabilization protocol, they will eventually adjust their successor and predecessor and the finger tables as well.

6.4.4 Performance

In a steady state, every node that participates in Chord holds information about $O(\log N)$ other nodes and can resolve lookups via $O(\log N)$ messages. As nodes join and leave the system, Chord can with high probability maintain lookups with no more than $O(\log^2 N)$ messages. A number of improvements have been developed for Chord—for example, for address space balancing and load balancing arbitrary item distributions across nodes [180].

6.5   Pastry

Pastry is a scalable, self-organizing, structured peer-to-peer overlay network [277]. Pastry is similar to Chord, CAN, and Tapestry. In this system, nodes and objects are assigned random identifiers from a large identifier space. Node identifiers (nodeIds) and keys are 128 bits long and can be seen as a sequence of digits in base 2^b, where b is a configuration parameter (typically 3 or 4).

Pastry builds on consistent hashing and the Plaxton's algorithm. It provides an object location and routing scheme and routes messages to nodes. It is a prefix-based routing system, in contrast to suffix-based routing systems such as Plaxton and Tapestry, that supports proximity and network locality awareness. At each routing hop, a message is forwarded to a numerically closer node. As with many other similar algorithms, Pastry uses an expected average of $\log(N)$ hops until a message reaches its destination.

Similar to the Plaxton's algorithm, Pastry routes a message to the node with the nodeId that is numerically closest to the given key. This node is called the key's root. Pastry nodes use three key functions internally to join the network and to route messages to nodes:

- *nodeId = pastryInit(Credentials)*: This function joins the current node to an existing Pastry network or starts a new network. The function returns a nodeId for the current node. The credentials argument can be used by the current node to provide authentication information to the network.

- *route(msg,key)*: This function routes a message (msg) to the node with nodeId numerically closest to the given key.
- *send(msg,IP-addr)*: This function sends a message to the node with the specific IP address if that node is available.

The Pastry system provides three key API calls for applications. Applications can use these function to implement additional features.

- *deliver(msg,key)*: This up-call is called by the Pastry when a message is received by the current node and the current node's nodeId is numerically the closest to the destination key of the message. This up-call is also used when a message has been explicitly sent to the current node's IP address.
- *forward(msg,key,nextId)*: This function is called by the Pastry before a message is forwarded to the node that has the identifier nextId. The application may inspect the message and modify its contents. The application can also stop the forwarding of the message by setting nextId to NULL.
- *newLeafs(leafSet)*. This function is called by the Pastry when there is a change in the leaf set (set of neighboring nodes). Applications can use this function to react to changes in their neighbor sets.

6.5.1 Joining and Leaving the Network

When a new node joins the network and needs to build its routing tables, this information is used to select nodes that are close to the new node. The leaf set contains nodes which are numerically close to the local node (both higher and lower). It is used as the first option if the destination node identifier is within the leaf range (numerically close to the current node); otherwise the prefix-based routing scheme is used.

A joining node needs to initialize its routing table and then inform other nodes that it has joined the network. The assignment of node identifiers is application-specific—typically it is computed using SHA-1 on the IP address or a public key. The joining node asks an existing node to route a special join message with the key equal to its new identifier. The message is routed to the existing node that is closest to the new node. Intermediate nodes receive this request, and they send their routing tables to the new node. The new node can then start to populate its routing table.

A Pastry node may fail or depart without prior warning. The Pastry network can handle such cases. A Pastry node fails when its immediate neighbors cannot communicate with it. A failed node needs to be replaced in the routing tables. This is accomplished by connecting the active node with the largest index on the side of the failed node. This node's routing table will have some nonoverlapping entries that are candidates to replace the failed node. One of them is selected as the replacement.

6.5.2 Routing

In order to route messages, each node maintains a routing table and a leaf set. A node's routing table has about $l = \lceil \log_{2^b} N \rceil$ rows and 2^b columns. The entries in row r of the routing table refer to nodes whose nodeIds share the first r digits with the local node's nodeId. The $(r + 1)$th nodeId digit of a node in column c of row r equals c. The column in row r corresponding to the value of the $(r + 1)$th digit of the local node's nodeId remains empty. At each routing step, a node normally forwards the message to a node whose nodeId shares with the key a prefix that is at least one digit longer than the prefix that the key shares with the present node's id. If no such node is known, the message is forwarded to a node

0	1	2	3	4	5	6	7	8	9	a	b	c	d	e	f
0X	1X	2X	3X	4X	5X		7X	8X	9X	aX	bX	cX	dX	eX	fX
60X	61X	62X	63X	64X			66X	67X	68X	69X	6aX	6bX	6cX	6dX	6eX	6fX
650X	651X	652X	653X	654X	655X	656X	657X	658X	659X		65bX	65cX	65dX	65eX	65fX
65a0X		65a2X	65a3X	65a4X	65a5X	65a6X	65a7X	65a8X	65a9X	65aaX	65abX	65acX	65adX	65aeX	65afX

Routing table of a Pastry node with nodeId
65a1x, $b = 4$. Digits are in base 16,
X represents an arbitrary suffix.

The IP address associated with each entry
is not shown.

FIGURE 6.4
Example of a Pastry routing table.

whose nodeId shares a prefix with the key, as long as the current node's nodeId exists but is numerically closer.

Each Pastry node maintains a set of neighboring nodes in the nodeId space (called the leaf set), both to ensure reliable message delivery and to store replicas of objects for fault tolerance. The expected number of routing hops is less than $\log_{2^b} N$. The Pastry overlay construction observes proximity in the underlying Internet. Each routing table entry is chosen to refer to a node with low network delay among all nodes with an appropriate nodeId prefix. As a result, one can show that Pastry routes have a low delay penalty: the average delay of Pastry messages is less than twice the IP delay between source and destination.

Figure 6.4 illustrates a Pastry routing table with a node that has the identifier *65a1x*, $b = 4$, and $l = 4$. The numbers are in base 16. In the addresses, "X" represents an arbitrary suffix. Each entry has an associated IP address that is not shown. Empty columns indicate digits corresponding to the present nodes' nodeId.

Figure 6.5 illustrates the prefix-based routing [65]. Node *65a1fc* routes a message to destination *d46a1c*. The message is routed to the nearest node in the identifier circle that is responsible for the address space of the destination.

Algorithm 6.3 presents pseudocode for the Pastry routing process. This algorithm is executed when a message with key D arrives at a node with nodeId A.

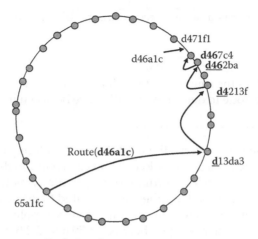

● Prefix-based
● Route to node with shared prefix
(with the key) of ID at least one
digit more than this node

FIGURE 6.5
Example of Pastry prefix routing.

6.5.3 Performance

Pastry requires $\log(N)$ hops until a message reaches its destination. With concurrent node failures, eventual delivery is guaranteed unless $l/2$ or more nodes with adjacent nodeIds fail simultaneously (l is an even integer parameter with typical value 16). The tables required in each Pastry node have $(2^b - 1) \times \lceil \log_{2^b} N \rceil + l$ entries. After a node failure or the arrival

---

Algorithm 6.3 Pseudocode for Pastry routing algorithm

Data: M is the neighborhood set, D is the destination address, A is the current node, L is the leaf set, L_i denotes the ith closest node identifier in the leaf set, and $R_{l,i}$ is the entry in the routing table R at column i, $0 \leq i \leq 2^b$ and row l, $0 \leq l \leq \lfloor 128/b \rfloor$

if $L_{-\lfloor |L|/2 \rfloor} \leq D \leq L_{\lfloor |L|/2 \rfloor}$ then

    /*D is within range of the leaf set or is the current node   */

    forward to an element L_i of the leaf set with GUID closest to D or the current node

else

    /*Use the routing table   */

    Let $l = shl(D, A)$

    /*l is the longest common prefix of D and A */

    if $R_{l,i} \neq null$ then

        forward to $R_{l,i}$

        /* forward to a node with a longer common prefix   */

    else

        /*There is no entry in the routing table   */

        forward to any node T in $L \cup R \cup M$ that has a common prefix of length l but is numerically closer

    end

end

---

of a new node, the invariants in all affected routing tables can be restored by exchanging $O(\log_{2^b} N)$ messages.

> The Pastry proximity metric is a scalar value that reflects the distance between any pair of nodes, such as the round-trip time. It is assumed that a function exists that allows each Pastry node to determine the distance between itself and a node with a given IP address.

The short routes property concerns the total distance, in terms of the proximity metric, that messages travel along Pastry routes. Recall that each entry in the node routing tables is chosen to refer to the nearest node according to the proximity metric with the appropriate nodeId prefix. As a result, in each step a message is routed to the nearest node with a longer prefix match. Simulations performed on several network topology models show that the average distance traveled by a message is between 1.59 and 2.2 times the distance between the source and destination in the underlying Internet.

Pastry has the *local route convergence* property: the routes of messages sent to the same key from nearby nodes in the underlying Internet tend to converge at a nearby intermediate node.

6.5.4 Bamboo

The Pastry's geometry and routing algorithm are used in another DHT algorithm called the Bamboo [271]. The main difference between the Pastry and Bamboo is that the latter improves routing table management to better handle churn. Resilience is increased by keeping more nodes in the leaf sets. Given that the routing table is incomplete, leaf sets can be used to allow progress in routing with the price of potentially longer paths. It has been shown by Gummadi et al. that, with a leaf set of 16 nodes, a random 30% link failure can be tolerated in a network of 65,536 nodes and still ensure that a path is found between two nodes [152].

6.6 Koorde

Koorde is a DHT based on Chord and the de Bruijn graphs. While inheriting the simplicity of Chord, Koorde introduces flexibility in balancing between optimal routing table size and average hop count in routing. Koorde embeds a deBruijn graph on the identifier circle for forwarding lookup requests. A node and a key have identifiers that are uniformly distributed in a 2^b identifier space. A key k is stored at its successor, the first node n that follows k on the identifier circle, whereas node $2^b - 1$ is followed by node 0. The successor of key k is identified as *successor(k)*.

6.6.1 Routing

Each node that joins the system m maintains knowledge about two other nodes: the node that succeeds it on the ring (its successor) and the first node, d, that precedes $2m$ (m's first de Bruijn node). Since the de Bruijn nodes follow each other directly on the ring, there is no reason to keep a variable for the second de Bruijn node ($2m + 1$); it is likely that d is also the predecessor for $2m + 1$.

A node has two outgoing edges: node m has an edge to node $2m \bmod 2^b$ and an edge to node $2m + 1 \bmod 2^b$. In other words, a node m points at the nodes identified by shifting a

---

Algorithm 6.4 Pseudocode for Koorde lookup function

Data: d contains the predecessor of $2m$. *Successor* contains the successor of m. k is the key, i is the imaginary de Bruijn node.

Function: $m.KoordeLookup(k,k_{shift},i)$

if $k \in (m, successor]$ **then**

| **return** *successor*

end

else if $i \in (m, successor]$ **then**

| $d.KoordeLookup(k, k_{shift} << 1, i \circ topBit(k_{shift}))$

end

return $successor.KordeLookup(k,k_{shift},i)$

---

new low-order bit into m and dropping the high-order bit. We represent these nodes using concatenation $\mod 2^b$, writing $m \circ 0 = 2m + \mod 2^b$ and $m \circ 1 = 2m + 1 \mod 2^b$.

As already mentioned in this chapter, routing a message from node m to node k in de Bruijn graph is accomplished by taking the number m and shifting in the bits of k one at a time until the number has been replaced by k. Each shift corresponds to a routing hop to the next intermediate address.

Algorithm 6.4 shows Koorde routing as an extension of the de Bruijn routing. Koorde passes the current imaginary node i as an argument to the routing function. In a single routing step, Koorde simulates the hop from imaginary node i to imaginary node $i \circ topBit(k)$, shifting in k. Koorde does so by hopping to $m.d$, which will have value near $2m$ and hopefully be equal to $predecessor(i \circ topBit(k))$. If so, Koorde iterates the next routing step. If at every hop, d is indeed the predecessor of $i \circ topBit(k)$, then Koorde contacts b nodes, where b is the number of bits in identifiers, because the algorithm shifts i left 1 bit at each hop.

Koorde's de Bruijn pointer is merely an important performance optimization; a query can always reach its destination slowly by following successors. Because of this property, Koorde can use Chord's join algorithm. Similarly, to keep the ring connected in the presence of nodes that leave, Koorde can use Chord's successor list and stabilization algorithm.

6.6.2 Performance

The optimal bound for a DHT is $O(\log n)$ hops using constant degree. The lower bound with a degree of $O(\log n)$ is $O((\log n)/\log\log n)$ hops. Koorde meets these bounds, such as $O(\log n)$ hops per lookup request with two neighbors per node (where n is the number of nodes in the DHT) and $O(\log n/\log\log n)$ hops per lookup request with $O(log n)$ neighbors per node [176].

6.7 Tapestry

Tapestry is a DHT system designed for best-effort routing to distributed objects in an efficient fashion. Tapestry is based on the Plaxton's algorithm and extends it with support for more reliability and churn. Figure 6.6 illustrates the Tapestry overlay network [362, 363].

In a similar fashion to Plaxton and Scribe, each routing table is organized in routing levels and each entry points to a set of nodes closest in network distance to a node that matches the suffix. In addition, a node also keeps back-pointers to each node referring to it. While Plaxton's algorithm keeps a mapping (pointer) to the closest copy of an object, Tapestry

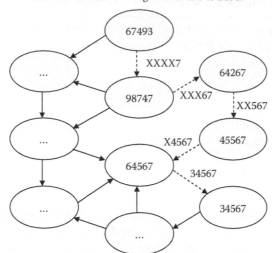

FIGURE 6.6
Example of a Tapestry network.

keeps pointers to all copies. This allows the definition of application-specific selectors about what object should be chosen (or what path).

Each node is assigned a unique node identifier uniformly distributed in the identifier space. The SHA-1 hash algorithm is used to create a 160-bit identifier space. Each identifier is represented using a 40-digit hexadecimal key. The assumption is that node identifiers and similarly constructed application identifiers are roughly evenly distributed over the identifier space. Typically, each node stores multiple application identifiers.

The Tapestry API includes the following functions:

- PublishObject, which is used to make an object (with a unique identifier) available
- UnPublishObject, which is used to remove an object from the network
- RouteToObject, which routes the given message to the destination object (exact match)
- RouteToNode, which routes a message to a node (closest match)

Tapestry provides an overlay routing network that aims to be stable under a variety of network conditions. Therefore, the system is an infrastructure for distributed applications and services. A number of prototype applications and services have been developed on top of Tapestry, including the following:

- OceanStore: Distributed storage utility on PlanetLab [191]
- Bayeux: Self organizing multicasting application
- Mnemosyne: A P2P steganographic file system [159]
- Spamwatch: Decentralized spam filter

6.7.1   Joining and Leaving the Network

The joining process is similar to the systems already covered. First, a new node needs to find an existing overlay node and an identifier for itself. Then the new node's routing table is constructed. After the new node has a routing table, it needs to inform other nodes of its presence so that they can route to it. The dynamic insertion of nodes therefore requires a moderate amount of processing. Node removal, however, is straightforward and easy.

6.7.2 Routing

Each identifier (object) is mapped to an active node called the root. A server S publishes that it has an object O by routing a message to the root of O using the overlay system in similar fashion to the Plaxton's algorithm, using incremental suffix routing. The original Plaxton scheme used the greatest number trailing bit positions to map an object to a node. In a distributed system there may be potentially many candidate nodes. Plaxton solved this using global ordering of nodes. Tapestry solves this by using a technique called *surrogate routing*. Surrogate routing tentatively assumes that an object's identifier is also the nodes identifier and routes a message using a deterministic selection toward that destination. The destination then becomes a surrogate root for the object.

When a query is sent to the identifier, it is routed toward the O's root. If a location mapping is encountered on the way at intermediate overlay nodes, the query is redirected to the server S. Tapestry allows overlay nodes to maintain multiple pointers per object, whereas Plaxton's algorithm only uses one. Applications can define the selection operator, and each object may include an optional application-specific metric. Objects can also have multiple roots, which helps to improve reliability.

Tapestry's routing table is very similar to Plaxton's, with additional information pertaining to the objects and back pointers. Each node maintains a list of back pointers, which point to nodes where it is referred to as a neighbor. These are used in dynamic node insertion algorithms to generate the routing tables for new nodes. Dynamic algorithms are employed for node insertion, populating routing tables, and notifying neighbors of new node insertions.

Figure 6.7 illustrates the Tapestry routing table at a node. Each neighbor map has multiple levels, where each level contains links to nodes matching up to a certain digit position in

Tapestry Routing Table

Entries \ Levels	1 Primary neighbor	2	3	4
1	0642	X042	XX02	XXX0
2	1642	X142	XX12	XXX1
3	2642	X242	XX22	XXX2
4	3642	X342	XX32	XXX3
5	4642	X442	XX42	XXX4
6	5642	X542	XX52	XXX5
7	6642	X642	XX62	XXX6
8	7642	X742	XX72	XXX7

Object location pointers
Hotspot monitor
Object store

Each entry can have multiple pointers for the same object. Objects can have multiple roots using salt value in hashing.

Back pointers

FIGURE 6.7
Example of a Tapestry routing table.

Algorithm 6.5 Pseudocode for Tapestry nextHop routing algorithm

---

Data: n is the previous hop number, G is the destination GUID, β is the base of the GUIDs (width of the routing table), $R_{i,j}$ is the routing table, in which the ith entry in the jth level is the ID and location of the closest node that begins with prefix(N, $j-1$)+i. *MaxHop(R)* is the height of R.

Function: *nextHop* returns the next hop or self if local node is the root

if $n = MaxHoP(R)$ **then**

> /*Destination reached   */
> **return self**

else

> $d = G_n$ /*d is the nth digit of G   */
>
> $e = R_{n,d}$ /*e is the dth entry of the nth row of R   */
>
> **while** $e = nil$ **do**
>
> > /*Incremental suffix routing   */
> >
> > $d = (d + 1) \bmod \beta$
> > $e = R_{n,d}$;
>
> **end**
>
> **if** $e = self$ **then**
>
> > **return** *NextHop($n+1,G$)*
>
> **else**
>
> > **return** e
>
> **end**

end

---

the ID. The primary ith entry in the jth level is the identifier and location of the closest node that ends with "i" + suffix(N, $j-1$). This means that level 1 has links to nodes that have nothing in common, level 2 has the first digit in common, and so on. The key difference to a routing table in the Plaxton overlay is that there can be multiple pointers per object and many roots per object. Tapestry also features algorithms for detecting hotspots and offers hints where additional replicas of objects need to be placed to alleviate load concerns.

Algorithm 6.5 presents the Tapestry's nextHop algorithm, which implements the incremental suffix routing using the routing table.

Tapestry provides fault tolerance by assigning multiple roots to each object. This is realized by adding a salt value to each object identifier to find alternate root identifiers. This effectively results in multiple identifiers for the same object. Soft-state lease is kept for each mapping entry in a routing table, and they are periodically refreshed and updated.

The Plaxton overlay assumes a static node population. Tapestry extends its design to adapt to the transient populations of P2P networks and provide adaptability, fault tolerance, and various optimizations. This soft state, using the announce/listen approach, is used to recover from failures in routing. In addition, the neighbor map is extended to maintain two backup neighbors in addition to the primary neighbor.

6.7.3   Performance

Experiments indicate that Tapestry efficiency increases with network size so that multiple applications sharing the same overlay network improves efficiency. Routing in Tapestry requires approximately $log_b N$ hops in a network of size N and identifiers of base b (hex-based digits have $b=16$). If an exact identifier cannot be found, the routing table will route to the closest matching node. For fault tolerance, nodes keep c secondary links such that the routing table has size $c \times b \times \log_B N$.

6.8 Kademlia

Kademlia is a scalable decentralized P2P system based on the XOR geometry (and metric) presented in Chapter 5 [182, 226, 311]. The algorithm is used by the BitTorrent DHT MainLine implementation, and therefore it is widely deployed. Kademlia is also used in kad, which is part of the eDonkey P2P file-sharing system that hosts several million simultaneous users. Relying on the XOR geometry makes Kademlia unique compared to other proposals built using rings, tori, butterflies, and similar geometries.

XOR was chosen as the geometry and metric for Kademlia because it has some useful properties in common with geometric distance:

- The XOR distance between a node and itself is zero.
- XOR is symmetric, and the distance from A to B and from B to A are equal.
- XOR satisfies the triangle inequality property.

Since XOR is symmetric, Kademlia peers can receive lookup queries from exactly the same distribution of nodes that are contained in their routing tables. Systems that do not have this symmetricity, such as Chord, do not learn useful routing information from queries being propagated [226].

> The Kademlia algorithm is based on the determination of the distance between two nodes. This distance is calculated as the *exclusive or* of the two input node identifiers. The result is taken as an integer number. This same scheme is used for calculating the distance between a node identifier and a key [226].

6.8.1 Joining and Leaving the Network

A node joining the Kademlia network undergoes a bootstrap process. The node needs to know the IP address and port of another node. If the bootstrapping node has not yet participated in the network, it computes a random and non-used identifier, which is used until the node leaves the network.

The initiating node maintains a shortlist of k closest nodes. These are probed to determine if they are active. The replies of the probes are used to improve the shortlist. Closer nodes replace more distant nodes in the shortlist. This iteration continues until k nodes have been successfully probed and these subsequent probes do not reveal improvements.

This process of locating k closest nodes to some node identifier is called a node lookup, and it is used in most operations offered by Kademlia. The procedure can be implemented either using recursively or iteratively. The current Kademlia implementation uses the iterative process where the control of the lookup is with the initiating node.

Leaving the network is straightforward and consistency is achieved by using leases.

6.8.2 Routing

In Kademlia, a node's neighbors are called *contacts*. They are stored in buckets, each of which holds a maximum of k contacts. These k contacts are used to improve redundancy.

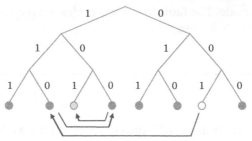

Every hop brings us in a smaller subtree around
the target and can forward requests to any
node in the appropriate subtree.

FIGURE 6.8
Kademlia XOR tree.

The routing table can be viewed as a binary tree in which each node in the tree is a k-bucket. Figure 6.8 presents an example XOR tree.

The buckets are organized by the distance between the current node and the contacts in the bucket. For the distance $d(n, c)$ between a node and its contact c in bucket j, where $0 \leq j < k$, the system has the invariant

$$2^j \leq d(n, c) < 2^{j+1}. \tag{6.2}$$

Every k bucket corresponds to a specific distance from the node. Nodes that are in the nth bucket must have a differing nth bit from the node's identifier. With an identifier of 128 bits, every node in the network will classify other nodes in one of 128 different distances.

Initially Kademlia nodes have only one k bucket. When the k bucket becomes full, it can be split. The split occurs if the range of nodes in the k bucket spans the nodes' own identifier.

When a node receives an update from another node, it updates the corresponding bucket. If the contact already exists, it is moved to the end of the bucket. Otherwise, if the bucket is not full, the new contact is added at the end.

Kademlia's routing table results in the same routing entries as for tree geometries when failures do not occur, such as Plaxton's algorithm. When failures occur, Kademlia can route around failures due to its geometry. Even though a message cannot be forwarded toward the destination with the highest differing bit fixed, it can make progress in the XOR distance toward the destination by fixing a lower-order bit.

Kademlia supports multiple paths between a source and a destination. These paths, however, may not have equal lengths. The XOR distance is separate from the network proximity. The geometry, however, has some flexibility in terms of what bits to fix and can thus be combined with network proximity awareness.

Figure 6.9 illustrates the lookup process. The first step is to inspect the client's routing table for the target identifier. A route is guaranteed when the high-order b bits match. The route points to another peer that is contacted in the next step. This peer is guaranteed to have a route when the first $2b$ bits match. This process continues until the closest peer to the identifier is found.

6.8.3 Performance

The key properties of Kademlia are prefix-based routing using the XOR-metric, redundancy in routing tables (k-buckets), parallel routing, and iterative routing [311].

The routing tables of all Kademlia nodes can be seen to collectively maintain one large binary tree. Each peer maintains a fraction $O(\log(n)/n)$ of this tree. During a lookup,

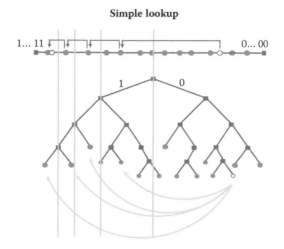

Simple lookup

FIGURE 6.9
Kademlia lookup.

each routing step takes the message closer to the destination requiring at most $O(\log n)$ steps.

6.9 Content Addressable Network

> The *content addressable network (CAN)* is a DHT algorithm based on virtual multi-dimensional Cartesian coordinate space [267]. In a similar fashion to other DHT algorithms, CAN is designed to be scalable, self-organizing, and fault tolerant. The algorithm is based on a d-dimensional torus that realizes a virtual logical addressing space independent of the physical network location. The coordinate space is dynamically partitioned into zones in such a way that each node is responsible for at least one distinct zone.

Each CAN node maintains a routing table that contains the IP address and virtual coordinate zone of its neighbors. Routing and forwarding using CAN is straightforward. A CAN node routes a message toward the message destination by choosing the closest neighbor to this point in the coordinate space. This involves first resolving the neighboring zone closest to the destination and then resolving the IP address of the neighbor responsible for that zone.

6.9.1 Joining the Network

In order for a new node to join the CAN network, the new node must first find a node that is already part of the network, identify a zone that can be split, and then update routing tables of neighbors to reflect the split introduced by the new node. In the seminal CAN article, the bootstrapping mechanism is not defined [267]. One possible scheme is to use a DNS lookup to find the IP address of a bootstrap node (essentially a rendezvous point). Bootstrapping nodes may be used to inform the new node of IP addresses of nodes currently in the CAN network.

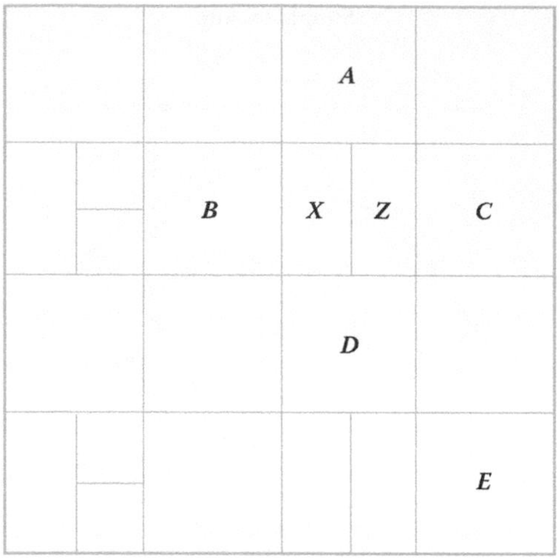

Peer X's coordinate neighbor set = {$A B D Z$}
New Peer Z's coordinate neighbor set = {$A C D X$}

FIGURE 6.10
Content addressable network.

After obtaining the IP address of an existing node, the new node can attempt to find a zone for itself. To find a zone, the new node selects a random point in the coordinate space and sends a join request with this point as the destination. The CAN overlay will route the message toward the destination using its routing tables. The message will eventually arrive at a node that is responsible for the zone to which the point belongs. This node may then choose to split the zone in half and give the second half to the new node. If the node responsible for the zone does not give up the zone, the new node needs to pick a new random point. After the zone split, the neighboring nodes are updated to reflect the two new zones and their IP addresses.

Figure 6.10 presents an example CAN network with a node X and its neighbors highlighted. The peer X initially had the neighbor set { A, B, C, D }. When a new peer Z is introduced, it will split X's coordinate space into two parts and Z will replace C as X's neighbor. The neighbor set of Z is { A, C, D, X }.

6.9.2 Leaving the Network

Node departures are handled in a fashion similar to joins. A node that is departing must give up its zone, and the CAN algorithm needs to merge this zone with an existing zone. The routing tables need to be updated then to reflect this change in zones. A node's departure can be detected using heartbeat messages that are periodically broadcast between neighbors. A node may also proactively indicate that it is leaving the network.

After a departing node has been identified, the next step for the CAN system is to merge the zone with an existing zone or take over the zone. The zone is tested for mergeability with existing zones. If a valid zone candidate is found among the neighbors, the zones are merged. If a merging candidate cannot be found, the neighboring node with the smallest zone will take over the departing node's zone. After the process, the neighboring nodes' routing tables are updated to reflect the change in the zone responsibility. The nodes may periodically attempt to merge additionally controlled zones with their neighbors.

Algorithm 6.6 Pythagorean-based CAN algorithm

Data: c is current node, P is the target point.
Function: *RouteCAN(c,P)* returns the owner p of point P
if $P \in c$ **then**
$\quad$ /* P is in c's neighbors n $\qquad\qquad\qquad\qquad\qquad\qquad$ */
$\quad$ $p \leftarrow n$
else
$\quad$ /*P is not in origo node c's zone $\qquad\qquad\qquad\qquad\qquad$ */
$\quad$ $p \leftarrow c$ /* current node is set to p $\qquad\qquad\qquad\qquad\qquad$ */
$\quad$ **while** $P \neq p$ **do**
$\quad\quad$ /* Until an owner is found for P $\qquad\qquad\qquad\qquad$ */
$\quad\quad$ $d \leftarrow \sqrt{((P_x - n_x)^2 + (P_y - n_y)^2}$
$\quad\quad$ Neighbor n with shortest distance d is the next hop node
$\quad\quad$ $p \leftarrow n$
$\quad$ **end**
end
Point P is in the current node p's zone
return p

6.9.3 Routing

A number of CAN routing algorithms have been proposed. These algorithms differ in how the routing decision is made. The routing algorithms can be partitioned as follows [260]:

- Pythagorean-based algorithm.
- Greedy forwarding along the x and y axes.
- Greedy forwarding with shortcut nodes.
- Inclination angle–based algorithms.
- Binary-based routing.

Recent results indicate that greedy forwarding algorithms perform better in a CAN network than the Pythagorean- and inclination angle–based algorithms [260]. For simplicity, we examine the Pythagorean-based algorithm (presented in Algorithm 6.6), which utilizes the Pythagorean theorem and calculates the shortest distance (hypotenuse) to the destination.

6.9.4 Performance

For a d-dimensional coordinate space partitioned into n zones, the average routing path length is $O(d \times N^{1/d})$ hops and each node needs to maintain $2d$ neighbors. This means that for a d-dimensional space the number of nodes can grow without increasing per node state. Another beneficial feature of CAN is that there are many paths between two points in the space, and thus the system may be able to route around faults. A logarithmic CAN is a system with $d = \log n$. In this case, CAN exhibits similar properties as Chord and Tapestry—for example, $O(\log n)$ diameter and degree at each node.

CAN supports proximity routing that does not require changes to routing table construction and maintenance. Network proximity is taken into account by measuring the *round-trip time (RTT)* between neighbors and forwarding messages to neighbors. This mechanism involves a trade-off between the number of hops in the path against the network distance traversed at each hop.

6.10   Viceroy

The Viceroy is a decentralized overlay algorithm that is designed to handle the discovery and location of data. The algorithm employs consistent hashing in a similar way to Chord; however, it uses a constant degree connection graph to achieve logarithmic diameter approximation to a butterfly network [217]. Figure 6.11 illustrates the butterfly network nature of Viceroy, which uses links between successors and predecessors on the ring for short distances. The basic idea of using shortcuts stems from Kleinberg's work on small worlds [186]. Barrière et al. [23] extended Kleinberg's work for rings instead of grids.

> The key point in Viceroy is the emphasis on constant degrees. The primary motivation was to develop an algorithm that has constant linkage cost, logarithmic path length, and best achievable congestion under the constraints. It generally has constant degree such as CAN. Its degree is smaller than in Chord, Tapestry, and Pastry. Viceroy assumes a global ordering on all the nodes in the system, which may make practical deployments in decentralized environments challenging.

Routing on Viceroy networks uses links between successors and predecessors on the ring for short distances. Ring construction is augmented with a constant number of long-range contacts chosen in such a way that a localized routing strategy produces short paths.

6.10.1   Joining the Network

Algorithm 6.7 introduces a new node into the Viceroy network. When a peer joins the network, it takes a random but permanent identity and selects its level within the network. Each peer needs to keep some state regarding the network—namely, the ring pointers

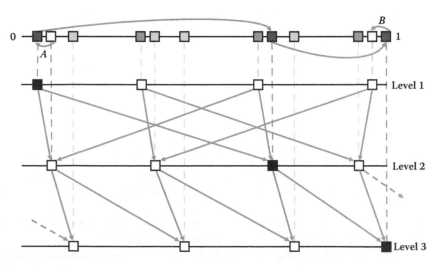

FIGURE 6.11
Request routing in Viceroy.

Algorithm 6.7 Viceroy join algorithm

Function: *joinViceroy()* joins a new node to the Viceroy network

1. Select identity *s*.
2. Each server picks its identifier independently and uniformly from [0, 1) and this identifier does not change while it is being used in the system.
3. Use lookup function to find *succ(s)*.
4. Update *predecessor* and *successor* of *s* and include *s* to the ring by inserting it into *pred(s):successor* and *succ(s):predecessor*.
5. Transfer all key-value pairs to *s* from the successor that are between *s:predecessor* and *s*.
6. Each node picks a level at random in such a way that when *n* servers are used, one of log *n* levels is chosen with nearly equal probability.
7. For a level *l* node, two edges are added connecting it to nodes at the level *l* + 1.
   - A *down-right* edge is added to a long-range contact at level *l* + 1 at a distance of approximately $1/2^l$.
   - A *down-left* edge is added at a close distance on the ring to level *l* + 1.
   - Furthermore, an *up* edge to a close-by node at level *l* − 1 is included if *l* > 1.
8. Finally, level-ring links (*nextonlevel, prevonlevel*) are added to the next and previous nodes of the same level *l*.

(*predecessor, successor*), the level ring pointers (*nextonlevel, prevonlevel*), and butterfly points (*down-left, down-right, up*). This adds up to seven links, which is the constant degree of the network.

6.10.2 Leaving the Network

When a peer departs, it passes its key pairs to a successor, and notifies other peers to find a replacement peer. Viceroy explicitly assumes that peers do not fail. It assumed that join and leave operations do not overlap in order to avoid the complication of concurrency mechanisms. In practice, both failures and concurrency support are required.

6.10.3 Routing

Algorithm 6.8 accepts the destination *x* and the current node *y* as input and finds the destination in the Viceroy network. The algorithm has three distinct stages.

Figure 6.11 presents an overview of a Viceroy network [213]. The figure illustrates a lookup from *A* to *B* using the dashed arrows. First, on the left side of the diagram, the lookup proceeds up from level 3 to level 1. The request is then forwarded to *down-right* to level 2. Finally, the request is forwarded to the destination at level 3 using the *down-right* link. The dark rectangles indicate nodes that process the request, and the nodes are also shown on the identifier space at the top of the diagram.

6.10.4 Performance

Viceroy and the work that motivated it can be said to be based on randomized routing topologies and to leverage the small-world phenomenon. This contrasts with the

---

Algorithm 6.8 Viceroy lookup algorithm

Function: *lookupViceroy(x,y)* finds the destination x from the current node y

1. First stage: The lookup starts with a climb using *up* connections to a level-1 node.
2. Second stage: Routing proceeds down the levels of the tree using the down links.
   - Moving from level l to level $l + 1$, the algorithm follows the *down-left* link if x is smaller than $1/2^l$.
   - Otherwise the *down-right* link is selected.
   - This continues until a node is reached with no *down* links, which is expected to be near the target.
3. Third stage: A vicinity search is conducted using the ring and level-ring links until the destination x is reached.

---

deterministic nature of Chord, Tapestry, Pastry, and CAN, which utilize the properties of the key identifier space to achieve scalability. Viceroy uses seven links and achieves $O(\log n)$ average latency. Kleinberg's proposal uses two links and achieves $O(\log n)^2$ average latency. Finally, the Symphony P2P system achieves $O((\log n)^2/k)$ average latency with $k + 1$ links.

6.11   Skip Graph

The final DHT algorithm to be discussed in this chapter is the skip graph. A skip graph is a probabilistic structure based on the skip list data structure [13, 160]. The skip list has simple and easy insert and delete operations that do not require tree rearrangements. Thus the operations are fast. The skip list is a set of layered ordered linked lists. All nodes are part of the bottom layer 0 list. Part of the nodes take part in layer 1 with some fixed probability. For each layer there is a probability for a node to be part of that layer. As a result, the upper layers of a skip list are sparse. This means that a lookup can quickly go through the list by traversing the sparse upper layer until it is close to the target. The downside of this approach is that the sparse upper layer nodes are potential hotspots and single points of failure. Skip graphs address this limitation and introduce multiple lists at each level to improve redundancy. Every node participates in one of the lists at each level. On average, $O(\log n)$ levels are needed in the structure, where n is the number of nodes.

The skip graph is a distributed version of the skip list, and its performance is comparable to the DHTs presented in this chapter. Each node in a skip graph has average of $\log n$ neighbors. The main benefit of the structure comes from its ability to support prefix and proximity search operations. DHTs guarantee that data can be located, but they do not typically guarantee where the data will be located. Skip graphs are able to support location-sensitive name searches because they use ordered lists [160].

In order to find a numeric object identifier, a skip graph–based search algorithm might search the lowest layer for the first digit and higher layers for the following digits. The ordered nature allows skip graphs to also support range searches. The algorithm can take network proximity into account so as to keep the search within an administrative boundary as far as possible. Although the structure has favorable properties (namely, support for range queries and flexibility in data placement), skip graph nodes require more links than

DHTs and thus result in increased maintenance traffic. The insert operation for a skip graph takes $O(\log n)$ time and messages. The search operation also takes $O(\log n)$ time and messages [13].

6.12 Comparison

Figure 6.12 summarizes the salient features of the DHT algorithms presented in this chapter. The algorithms are compared in terms of the following six features: foundation, routing function, system parameters, routing performance, routing state, and how join/leaves (churn) is handled.

System parameters differ as well based on the underlying foundation. The key parameter is the number of nodes N in the system. CAN has an additional dimension parameter d, Chord and Koorde use base 2, and some systems, such as Kademlia, Pastry, and Tapestry, have a base B. In general, the system achieves logarithmic routing performance, with the exception of CAN, which has a higher routing performance but constant routing state.

In this chapter, we observed that Pastry and Tapestry are based on the Plaxton's algorithm [257]. This algorithm was proposed in 1997 to improve web caching performance. The Plaxton's algorithm uses suffix routing to obtain delivery time within a small factor of the optimal delivery time. This algorithm is not suitable for decentralized and dynamic environments because it requires global knowledge and does not support additions and removals of nodes.

	CAN	Chord	Kademlia	Koorde	Pastry	Tapestry	Viceroy
Foundation	Multi-dimensional space (dimensional torus)	Circular space (hyper-cube)	XOR metric	de Bruijn graph	Plaxton-style mesh (hyper-cube)	Plaxton-style mesh (hyper-cube)	Butterfly network
Routing function	Maps (key, value) pairs to coordinate space	Matching key and nodeID	Matching key and nodeID	Matching key and nodeID	Matching key and prefix in nodeID	Suffix matching	Routing using levels of tree, vicinity search
System parameters	Number of peers N, number of dimensions d	Number of peers N	Number of peers N, base of peer identifier B	Number of peers N	Number of peers N, base of peer identifier B	Number of peers N, base of peer identifier B	Number of peers N
Routing performance	$O(dN^{1/d})$	$O(\log N)$	$O(\log_B N) +$ small constant	Between $O(\log \log N)$ and $O(\log N)$, depending on state	$O(\log_B N)$	$O(\log_B N)$	$O(\log N)$
Routing state	$2d$	$\log N$	$B \log_B N + B$	From constant to $\log N$	$2B \log_B N$	$\log_B N$	Constant
Joins/leaves	$2d$	$(\log N)^2$	$\log_B N +$ small constant	$\log N$	$\log_B N$	$\log_B N$	$\log N$

FIGURE 6.12
Comparison of DHT algorithms.

6.12.1   Geometries

We observe that the foundations differ across the algorithms but result in similar scalability properties. The foundations were considered earlier in the previous chapter, and for the considered systems the foundations are tori, ring, XOR metric, de Bruijn graph, hypercube, and butterfly network. The conclusions of several comparisons of the geometries are that the ring, XOR, and de Bruijn geometries are more flexible than the others and permit the choice of neighbors and alternative routes [152].

6.12.2   Routing Function

The routing function then utilizes the properties of the foundation in order to maintain routing tables and forward messages toward their destination. The routing function differs based on the algorithm and typically maps a key (that defines the destination) to a neighbor closer in the routing space.

The routing tables of DHTs can vary from size $O(1)$ to $O(n)$. The algorithms need to balance between maintenance cost and lookup cost. From the viewpoint of routing state, Chord, Pastry, and Tapestry offer logarithmic routing table sizes, whereas Koorde and Viceroy support constant or near-constant sizes. Churn and dynamic peers can also be supported with logarithmic cost in some of the systems, such as Koorde, Pastry, Tapestry, and Viceroy. Recent analysis indicates that large routing tables actually lead to both low traffic and low lookup hops. These good design points translate into one-hop routing for systems of medium size and two-hop routing for large systems [317].

The basic Plaxton scheme was the starting point for many DHT algorithms; however, it suffers from several limitations. The Plaxton scheme uses only one root node that is a single point of failure. Moreover, it does not allow nodes to be inserted and removed and assumes a total ordering of nodes. Both Tapestry and Pastry address these limitations. Tapestry uses surrogate routing to be able to incrementally choose root nodes. Tapestry addresses congestion by placing replicas close to nodes that generate high request loads. Tapestry allows nodes to select from a set of replicas, whereas the Plaxton scheme knows only the nearest replicas. Routing faults and node failures are detected by Tapestry using TCP timeouts and UDP heartbeats. Zhuang et al. have investigated a number of keep-alive algorithms for DHT failure detection [366].

6.12.3   Churn

Li et al. provide a comparison of different DHTs under churn [202]. They examine the fundamental design choices of systems, including Tapestry, Chord, and Kademlia. The insights based on this work include the following:

- Larger routing tables are more cost-effective than more frequent periodic stabilization.
- Knowledge about new nodes during lookups may allow elimination of the need for stabilization.
- Parallel lookups result in reduced latency due to timeouts, which provide information about the network conditions.

6.12.4   Asymptotic Trade-offs

Figure 6.13 illustrates the asymptotic trade-off curve between the routing table size and the network diameter [350]. We observe that there is a clear relation between these two metrics.

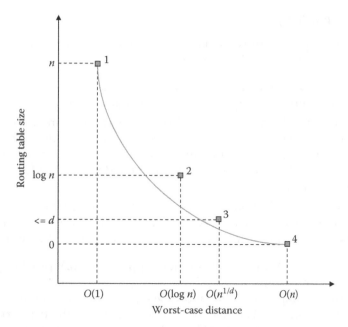

FIGURE 6.13
Asymptotic trade-off curve between routing table size and network diameter.

The extreme cases are illustrated in the figure at the edges of the graph (1 and 4), in which nodes either maintain full state resulting in diameter of 1, or they do not maintain any state, resulting in a system-wide broadcast with diameter n. Consistent hashing is an example of a technique that can achieve $O(1)$ lookup cost with $O(n)$ state. These two approaches are not desirable in practice. The former results in a full centralized directory that needs to be maintained, and the latter results in flooding that burdens the network and has long delays.

The intermediate cases are interesting. A large number of the presented overlay algorithms, such as Chord, Pastry, Tapestry, fall into the middle portion of the graph, with logarithmic diameter and routing table size (2). With constant routing performance, CAN requires somewhat larger routing table sizes (3).

Analysis of the trade-offs between the two metrics indicates that the routing table size of $\Omega(\log n)$ is a threshold point that separates two distinct state-efficiency regions. One can observe that this point is in the middle of the symbolic asymptotic curve. If the routing table size is asymptotically smaller or equal, the requirement for congestion-free operation prevents it from achieving the smaller asymptotic diameter. When the routing table size is larger, the requirement for congestion-free operation does not limit the system anymore.

A number of $O(1)$ lookup-cost DHT algorithms, also called one-hop DHTs, have been proposed (for example, Beehive [266] and Kelips [157]). The aim of these DHTs is to use more state and thus allow constant-time or near constant-time lookups. Beehive is a proactive replication scheme that runs on top of an $O(\log n)$ DHT algorithm. The system achieves $O(1)$ for queries following the power-law distribution. The system is motivated by the observations that DNS and Web requests follow power-law distributions. A fast random sampling technique has also been proposed for DHT systems that uses geometric routing. This scheme achieves an average lookup latency comparable to the average unicast latency, given that the graph follows a power-law latency expansion [358].

6.12.5 Network Proximity

Support for network proximity is one key feature of overlay algorithms. The three basic models for proximity awareness in DHTs are [63]

- Geographic layout: Node identifiers are created in such a way that nodes that are close in the network topology are close in the nodeId space.

- Proximity routing: The routing tables do not take network proximity into account; however, the routing algorithm can choose a node from the routing table that is closest in terms of network proximity. At each hop, the chosen node strikes a balance between taking the message closer to the destination identifier and using the nearest neighbor.

- Proximity neighbor selection: In this model, the routing table construction takes network proximity into account. Routing table entries are chosen in such a way that at least some of them are close in the network topology to the current node.

The basic Chord and CAN protocols do not support network proximity; however, geographic layout and proximity routing have been proposed for CAN. CAN follows the network proximity routing model and allows geographically nearby nodes to be close in the identifier space as well. This poses some challenges when the network evolves differently in different places. Proximity neighbor selection is used with prefix- and suffix-based algorithms such as Pastry and Tapestry, and it has been found to be highly effective compared to the other models [63].

Proximity neighbor selection was found to yield significantly better paths than proximity routing. Moreover, the effectiveness of the proximity methods does not depend on the routing geometry. Kademlia's XOR routing [182] and Chord's ring geometry support proximity neighbor selection better than hypercube, which supports only proximity routing [152].

6.12.6 Adding Hierarchy to DHTs

Most DHTs that have been proposed are flat and nonhierarchical structures. They thus contrast with the traditional distributed systems, which have employed hierarchy to achieve scalability. A hierarchical DHT can be constructed that retains the homogeneity of load and functionality of the flat DHTs. A generic construction called Canon has been shown to offer the same routing state and routing hops trade-off found in the flat DHT designs [141]. The benefits of this approach include fault isolation, adaptation to the underlying physical network and its organizational boundaries, and hierarchical storage of content and access control.

The system is based on multiple domains, each running its own flat DHT algorithm. These DHTs can then be merged to form a structure that spans the multiple domains in a hierarchical fashion. Connectivity between domains is achieved by creating links between the domains. The challenge is to create the links in such a way that the routing state and average hop count are comparable to a single large DHT. The Canon design can be applied for many different flat DHT structures and different routing geometries. It has been demonstrated using DHTs such as Chord, CAN, and Kademlia.

We briefly examine how the Canon approach can be applied to the Chord DHT. Each node in this hierarchical variant of Chord, called Crescendo, is assigned a unique identifier from the circular identifier space. The link structure is recursive in nature, and each set of nodes in a leaf domain forms a Chord ring. At each internal domain, the Crescendo ring is formed by merging all the children Crescendo rings into a single Crescendo ring. This construction is repeated recursively at higher levels of the hierarchy. This process results in a global DHT that contains all the domains and nodes in the system [141].

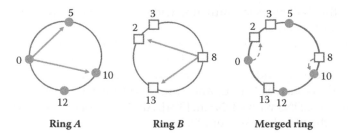

FIGURE 6.14
Merging Chord rings.

Figure 6.14 illustrates the merging of two Chord rings, A and B, into a Crescendo ring. Each of the rings has four nodes that have unique identifiers in the range $[0, 16]$. This example focuses on the edges created by two nodes, node 0 in ring A and node 8 in ring B. Based on the Chord algorithm, node 0 establishes its links in ring A by finding, for each $0 \leq k < 4$, the closest node that is at least distance 2^k. Thus the node has links to nodes 5 and to 10. In a similar fashion, node 8 in ring B has links to nodes 13 and 2.

In the ring merging operation, the nodes keep their original links. In addition, each node m in one ring creates a link to a node m' in the other ring if and only if

- m' is the closest node that is at least distance 2^k away for some $0 \leq k \leq N$,
- m' is closer to m than any node in the ring of m.

Based on these conditions, in the example node 0 links to node 2 in the other ring. Node 8 in ring B links to node 10 in ring A. Some nodes may not create additional links. This approach of merging two rings generalizes to the merging of any number of rings. The algorithm for link creation is applied bottom-up on the hierarchy by merging sibling rings to construct a level higher and larger ring.

Routing in Crescendo is identical to routing in the Chord DHT, namely greedy clockwise routing. When there are multiple levels of hierarchy, greedy clockwise routing takes the message to the closest predecessor p of the destination at each level. This node p is then responsible for taking the message to the next higher Crescendo ring.

6.12.7 Experimenting with Overlays

Overlay algorithms are typically evaluated using experimental measurements (traces) from real-life systems, simulations, and analytical models. The distributed PlanetLab testbed is frequently used to experiment with various overlay algorithms. All PlanetLab machines run a common software package that includes bootstrapping and management modules. PlanetLab supports distributed virtualization through a technique called slicing. Users can request PlanetLab slices in which they can run experiments and services. Thee services include file sharing and network-embedded storage, routing and multicast overlays, QoS overlays, event dissemination, anomaly-detection mechanisms, and network measurement systems [329].

There are many simulators available that support both structured and unstructured overlay algorithms. As a classic example we can take the ns-2 simulator,[1] which is a discrete event simulation framework that is commonly used to experiment with different TCP/IP

---

[1] The Network Simulator ns-2: http://www.isi.edu/nsnam/ns.

protocols. One of the challenges in simulating wide-area networks with ns-2 are the memory and processing requirements.

OMNeT++[2] is a component-based modular simulator architecture for experimenting with various networks. OverSim is an OMNeT++-based open-source simulation framework designed for overlay and P2P networks. This simulator supports a number of structured DHT algorithms such as Chord, Kademlia, and Pastry [25]. OMNeT++ also has been extended to simulate CDNs with CDNSim [304]. Another toolkit for overlay networks with modular design is the Overlay Weaver [292].

6.12.8   Criticism

There have been two main criticisms of structured systems [70]. The first pertains to peer transience, which is an important factor in maintaining robustness. Highly transient peers may not be well supported by DHTs [70]. Transient peers result in churn, which is a current concern with DHTs. The tolerance to churn depends on the DHT algorithm as well as the application scenario [201].

The second criticism of structured systems stems from their foundation in consistent hashing, which makes it more challenging to implement scalable query processing than for unstructured systems. Given that the popular file-sharing applications rely extensively on metadata-based queries, simple exact-match key searches are not sufficient for them, and additional solutions are needed on top of the basic DHT API. It is also possible to combine structured and unstructured algorithms in so-called hybrid models [61]. Unstructured networks with flooding or random walks are inefficient with sparse data; that is, data that is not widely replicated and available. In these environments, it is tempting to utilize a more structured approach to find keys efficiently irrespective of the level of replication. Castro, et al. have proposed Structella, which is a hybrid of Gnutella built on top of Pastry [61]. Another proposal employed structured search for rare items and unstructured search for massively popular and replicated items [212].

---

[2] http://www.omnetpp.org/

7

Probabilistic Algorithms

Many peer-to-peer (P2P) protocols and overlay networks utilize probabilistic techniques to reduce processing and networking costs. This chapter presents a number of frequently used and useful probabilistic techniques. Bloom filters and their variants are of prime importance, and they are heavily used in various network solutions.

The chapter also examines epidemic algorithms and gossiping, which are also the foundation of a number of overlay solutions. In basic gossip-based protocols, each node contacts a subset of nodes in each round and exchanges information with these nodes. The dynamics of information dissemination bear semblance to the spread of an epidemic [108] and can result in high robustness, reliability, and self-stabilization [31].

7.1 Overview of Bloom Filters

Fast matching of arbitrary identifiers to values is a basic requirement for a large number of applications. Data objects are typically referenced using locally or globally unique identifiers. Recently, many distributed systems have been developed using probabilistic globally unique random bitstrings as node identifiers. For example, a node tracks a large number of peers that advertise files or parts of files. Fast mapping from host identifiers to object identifiers and vice versa are needed. The number of these identifiers in memory may be great, which motivates the development of fast and compact matching algorithms.

Given that there are millions or even billions of data elements, efficient solutions for storing, updating, and querying them becomes increasingly important. The key idea behind the data structures discussed in this chapter is that, by allowing the representation of the set of elements to lose some information, in other words to become lossy, the storage requirements can be significantly reduced.

The data structures presented in this chapter for probabilistic representation of sets are based on the seminal work by Burton Bloom in 1970. Bloom first described a compact probabilistic data structure that was used to represent words in a dictionary. There was little interest in using Bloom filters for networking until 1995, after which this area has gained widespread interest both in academia and in the industry.

Bloom filters are an efficient mechanism for probabilistic representation of sets and support membership queries [32]. Bloom filters have many applications in dictionaries, networking, measurement, and P2P systems [40]. Meta-databases are an example application domain of Bloomier filters. Meta-databases direct queries to actual external databases.

Toward the end of the chapter, we consider four types of applications pertaining to distributed operation and networking: caching, P2P networks, packet routing and forwarding, and measurement.

7.2   Bloom Filters

The Bloom filter is a space-efficient probabilistic data structure that supports set membership queries. The data structure was conceived by Burton H. Bloom in the 1970s. The structure offers a compact probabilistic way to represent a set that can result in false positives but never in false negatives. This makes Bloom filters useful for many different kinds of tasks that involve lists and sets. The basic operations involve adding elements to the set and querying for element membership in the probabilistic set representation.

The basic Bloom filter does not support the removal of elements; however, a number of extensions have been developed that also support removals. The accuracy of a Bloom filter depends on the size of the filter, the number of hash functions used in the filter, and the number of elements added to the set. When more elements are added to a Bloom filter, the probability that the query operation reports false positives becomes higher.

> Broder and Mitzenmacher have coined the *Bloom filter principle* [40]:
>
> Whenever a list or set is used, and space is at a premium, consider using a Bloom filter if the effect of false positives can be mitigated.

A Bloom filter is an array of m bits for representing a set $S = \{x_1, x_2, \ldots, x_n\}$ of n elements. Initially all the bits in the filter are set to zero. The key idea is to use hash functions to map items in the set S to a random number uniform in the range $1, \ldots m$. A Bloom filter uses k hash functions and they are assumed to be random. The MD5 hash algorithm is a popular choice for the hash functions.

An element $x \in S$ is inserted into the filter by setting the bits $h_i(x)$ to one for $1 \leq i \leq k$. If the bits are not set, then x is not an element of S. If all the bits are set to one, then it is assumed that the element is a member of S. Algorithm 7.1 presents the pseudocode for the insertion operation. Algorithm 7.2 gives the pseudocode for the membership test of a given element x in the filter. The weak point of Bloom filters is the possibility for a false positive. False positives are elements that are not part of S but are reported being in the set by the filter.

Figure 7.1 presents an overview of a Bloom filter. The Bloom filter consists of a bitstring of length 18. Three elements have been inserted, namely x, y, and z. Each of the elements have

---

Algorithm 7.1 Pseudocode for Bloom filter insertion

Data: x is the object key to insert into the Bloom filter.
Function: *insert(x)*
for $j : 1 \ldots k$ **do**

    /* Loop all hash functions k */

    $i \leftarrow h_j(x)$
    if $B_i == 0$ **then**
        /* Bloom filter had zero bit at position i */
        $B_i \leftarrow 1$
    end
end

Algorithm 7.2 Pseudocode for Bloom member test

Data: x is the object key for which membership is tested.
Function: *ismember(x)* returns true or false to the membership test
$m \leftarrow 1$
$j \leftarrow 1$
While $m == 1$ and $j \leq k$ **do**

    $i \leftarrow h_j(x)$
    if $B_i == 0$ then
        $m \leftarrow 0$
        $j \leftarrow j + 1$
    end
end
return m

---

been hashed using three hash functions to bit positions in the bitstring. The corresponding bits have been set to 1. Now, when an element not in the set, w, is looked up, it will be hashed using the three hash functions into bit positions. In this case, one of the positions is zero and hence the Bloom filter reports correctly that the element is not in the set. It may happen that all the bit positions of an element report that the corresponding bits have been set. When this occurs, the Bloom filter will erroneously report that the element is a member of the set. These erroneous reports are called false positives. We observe that for the inserted elements, the hashed positions correctly report that the bit is set in the bitstring.

For optimal performance, each of the k hash functions should be a member of the class of universal hash functions, which means that the hash functions map each item in the universe to a random number uniform over the range. The development of uniform hashing techniques has been an active area of research. An almost ideal solution for uniform hashing is presented in [246]. In practice, reasonable hash functions, such as MD5, appear to be useful for most purposes. In addition, d-left hashing has been proposed as almost perfect hash function [38].

> A Bloom filter constructed based on S requires space $O(n)$ and can answer membership queries in $O(1)$ time. Due to its probabilistic nature, the structure has one-sided error. Given $x \in S$, the Bloom filter will always report that x belongs to S, but given $y \notin S$ the Bloom filter may report that $y \in S$.

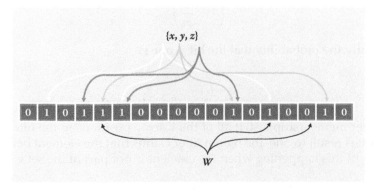

FIGURE 7.1
Overview of a Bloom filter.

	Decrease	Increase
Number of hash functions (k)	Less computation Higher false positive rate	More computation Lower false positive rate
Size of filter (m)	Smaller space requirements Higher false positive rate	More space is needed Lower false positive rate
Number of elements in the inserted set (n)	Lower false positive rate	Higher false positive rate

FIGURE 7.2
Key Bloom filter parameters.

Figure 7.2 examines the behavior of three key parameters when their value is either decreased or increased. The number of hash function is used to tune the accuracy of the filter with the price of more computation in insertions and lookups. The cost is directly proportional to the number of hash functions. The size of the filter can be used to tune the space requirements and the false positive rate. A larger filter will result in fewer false positives. Finally, the size of the set that is inserted into the filter determines the false positive rate.

7.2.1 False Positive Probability

Now, we derive the false positive probability rate of a Bloom filter and the optimal number of hash functions for a given false positive probability rate. We start with the assumption that a hash function selects each array position with equal probability. Let m denote the number of bits in the Bloom filter. When inserting an element to the filter, the probability that a certain bit is not set to one by a hash function is given by

$$1 - \frac{1}{m}. \tag{7.1}$$

Now, there are k hash functions, and the probability that any of them have not set the bit to one is given by

$$\left(1 - \frac{1}{m}\right)^k. \tag{7.2}$$

After inserting n elements to the filter, the probability that a given bit is still zero is

$$\left(1 - \frac{1}{m}\right)^{kn}. \tag{7.3}$$

And consequently the probability that the bit is one is

$$1 - \left(1 - \frac{1}{m}\right)^{kn}. \tag{7.4}$$

For an element membership test, if all of the k array positions in the filter computed by the hash functions result to one, the Bloom filter claims that the element belongs to the set. The probability of this happening when the element is not part of the set is given by

$$\left(1 - \left(1 - \frac{1}{m}\right)^{kn}\right)^k \approx \left(1 - e^{-kn/m}\right)^k. \tag{7.5}$$

We note that $e^{-kn/m}$ is a very close approximation of $(1 - (1/m))^{kn}$ [40]. The false positive probability decreases as the size of the Bloom filter increases (m). The probability increases as more elements are added to the filter and n increases. Now, we want to minimize the probability for false positives. Minimizing the false positive rate is performed by minimizing $(1 - e^{-kn/m})^k$ with respect to k. This is accomplished by taking the derivative. The minimal value of k is given by

$$\frac{m}{n} \ln 2 \approx \frac{9m}{13n}. \tag{7.6}$$

This results in the false probability of

$$\left(\frac{1}{2}\right)^k \approx 0.6185^{m/n}. \tag{7.7}$$

Taking the optimal number of hashes, the false positive probability (when ≤ 0.5) can be rewritten and bounded

$$\frac{m}{n} \geq \frac{1}{\ln 2}. \tag{7.8}$$

This means that in order to maintain a fixed false positive probability, the length of a Bloom filter must grow linearly with the number of elements inserted in the filter.

Figure 7.3 presents the false probability rate p as a function of number of elements n in the filter and the filter size m. An optimal number of hash functions $k = (m/n) \ln 2$ has been assumed.

There is a factor of $\log_2 e \approx 1.44$ between the amount of space used by a Bloom filter and the optimal amount of space that can be used. There are other data structures that use space closer to the lower bound, but they are more complicated.

7.2.2 Operations

Standard Bloom filters do not support the removal of elements. Removal of an element can be implemented by using a second Bloom filter that contains elements that have been

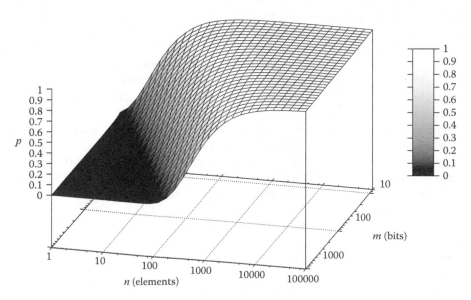

FIGURE 7.3
False probability rate for Bloom filters.

removed. The problem of this approach is that the false positives of the second filter result in false negatives in the composite filter, which is undesirable. Therefore, a number of dedicated structures have been proposed that support deletions. These are examined later in this chapter.

A number of operations involving Bloom filters can be implemented easily—for example, the union and halving of a Bloom filter. The bit-vector nature of Bloom filter allows the merging of two or more Bloom filters simply by performing bitwise OR on the bit-vector. Given two sets S_1 and S_2, a Bloom filter B that represents the union $S = S_1 \cup S_2$ can be created by taking the OR of the original Bloom filters $B = B_1 \vee B_2$. The merged filter B will report any element belonging to S_1 or S_2 as belonging to set S.

A Bloom filter can be halved easily in size. Given that the size of the filter is a power of two, the size can be halved by taking OR of the first and second halves together. The highest-order bit can be masked when hashing in lookups.

Bloom filters can be used to approximate set intersection; however, this is more complicated than the union operation. The inner product of the bit-vectors is an indicator of the size of the intersection [40]. The idea of a bloomjoin was presented by Mackert and Lohman in 1986 [216]. Bloomjoin is by two hosts, A and B, compute the intersection of two sets S_1 and S_2, when A has the first set and B the second. It is not feasible to send all the elements from A to B, and vice versa. In a bloomjoin, the S_1 is represented using a Bloom filter and sent from A to B. B can then compute the intersection and send back this set. Host A can then check false positives with B in a final round.

7.2.3   d-left Counting Bloom Filter

> Bonomi et al. [38] presented a data structure based on *d-left hashing* and fingerprints that is functionally equivalent to a counting Bloom filter but uses approximately half the space.

The d-left hashing scheme divides a hashtable into d subtables that are of equal size. Each subtable has n/d buckets, where n is the total number of buckets. When an element is placed into the table, hashing is used to obtain d possible buckets where it can be placed. The candidate buckets are obtained by applying independent uniform hash functions. Each incoming element is placed in the bucket that contains the smallest number of elements. In case of a tie, the element is placed in the bucket of the leftmost subtable with the smallest number of elements. When searching for an element, each of the d possible subtables are examined. The technique uses multiple choices in hashing to achieve balanced loads and very small maximum loads.

The d-left counting Bloom filter partitions the m bits among the k hash functions and creates k slices of $m' = m/k$ bits. Each hash function $h_i()$, with $1 \leq i \leq k$, results in an index over m' for the slice it is responsible for. When an element is inserted into the structure, it is first given a fingerprint. The fingerprint is stored in the least loaded subtable. Lookups use parallel search of the subtables to find the fingerprint.

When testing for an element x, if all k bits given by $h_i(x)$, $1 \leq i \leq k$ are set to one, the filter will result in a false positive. This false positive probability can be reduced by increasing the number of slices k or the size m. The problem of knowing which element to delete is solved by breaking the problem into two steps—namely, the creation of the fingerprint and then finding the k locations.

This approach results in a more robust filter that spreads the load more uniformly over the bits. Thus no element is specially sensitive to false positives [41].

7.2.4 Compressed Bloom Filter

> Compressing Bloom filter improves performance when a Bloom filter is passed in a message between distributed nodes. This structure is particularly useful when information must be transmitted repeatedly and the bandwidth is a limiting factor [230].

Compressed Bloom filters are used only for optimizing the transmission (over the network) size of the filters. This is motivated by applications such as Web caches and P2P information sharing, which frequently use Bloom filters to distribute routing tables. If the optimal value of the number of hash functions k in order to minimize the false probability is used then the probability that a bit is set in the bitstring representing the filter is 1/2. Given the assumption of independent random hash functions, this means that the bitstring is random, and thus it does not compress well.

The key idea in compressed Bloom filters is that by changing the way bits are distributed in the filter, it can be compressed for transmission purposes. This is achieved by choosing the number of hash functions k in such a way that the entries in the m vector have a smaller probability than 1/2 of being set. After transmission, the filter is decompressed for use. The size of k selected for compression is not optimal for the uncompressed Bloom filter, but may result in a smaller compressed filter. Compression can result in a smaller false positive rate as a function of the compressed size compared to a Bloom filter that does not use compression. The compressed Bloom filter requires that some additional compression algorithm is used for the data that is transmitted over the network, for example, arithmetic coding [230].

7.2.5 Counting Bloom Filters

As mentioned with the treatment on standard Bloom filters, they do not support element deletions. A Bloom filter can be easily extended to support deletions by adding a counter for each element of the data structure. This means that instead of having m bits we have m counters. Fan et al. [127] first introduced the idea of a counting Bloom filter in conjunction with Web caches.

Figure 7.4 illustrates a counting Bloom filter. The structure works in a manner similar to a regular Bloom filter; however, it is able to keep track of the insertions and deletions. In this example, three elements are added as follows. Element x is inserted three times, and y and z are inserted once. Three hash functions are used to find the bit positions for each element, and the corresponding counter in the filter is incremented by one.

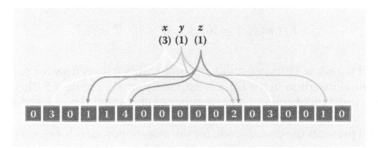

FIGURE 7.4
Example of a counting Bloom filter.

A counting Bloom filter also has the ability to keep a approximate counts of items. This count estimate can be determined by finding the minimum of the counts in all locations in the bitstring where an item is hashed to.

> In a counting Bloom filter, each entry in the Bloom filter is not a single bit but rather a small counter. When an item is inserted, the corresponding counters are incremented; when an item is deleted, the corresponding counters are decremented. To avoid counter overflow, we choose sufficiently large counters.

Given that the counters are only used for book-keeping for the membership test, the analysis from [127] reveals that 4 bits per counter should suffice for most applications [40, 128]. To determine a good counter size, we can consider a counting Bloom filter for a set with n elements, k hash functions, and m counters. Let $c(i)$ be the count associated with the ith counter. The probability that the ith counter is incremented j times is a binomial random variable:

$$P(c(i) = j) = \binom{nk}{j} \left(\frac{1}{m}\right)^j \left(1 - \frac{1}{m}\right)^{nk-j} \tag{7.9}$$

The probability that any counter is at least j is bounded above by $mP(c(i) = j)$, which can be calculated using the above formula.

The counter counts the number of times that the bit is set to one. All the counts are initially zero. The probability that any count is greater or equal to j:

$$Pr(max(c) \geq j) \leq m \binom{nk}{j} \frac{1}{m^j} \leq m \left(\frac{enk}{jm}\right)^j . \tag{7.10}$$

As already mentioned the optimum value for k (over reals) is $\ln 2m/n$ so assuming that the number of hash functions is less than $\ln 2m/n$ we can further bound

$$Pr(max(c) \geq j) \leq m \left(\frac{e \ln 2}{j}\right)^j . \tag{7.11}$$

Hence taking $j = 16$ we obtain that

$$Pr(max(c) \geq 16) \leq 1.37 \times 10^{-15} \times m. \tag{7.12}$$

In other words if we allow 4 bits per count, the probability of overflow for practical values of m during the initial insertion in the filter is extremely small. Figure 7.5 illustrates overflow probability as a function of counter size (number of elements). The case of 4 bit counters is shown using a horizontal line.

Algorithm 7.3 presents the pseudocode for the insert operation for element x with counting. The operation increments the counter of each bit to which x is hashed. The counting structure supports the removal of elements using the delete operation presented in Algorithm 7.4. The delete decrements the counter of each bit to which x is hashed. The corresponding bit is reset to zero when the counter becomes zero.

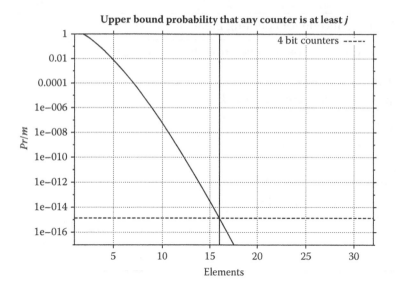

FIGURE 7.5
Counting Bloom filter scalability.

7.2.6 Hierarchical Bloom Filters

Shanmugasundaram et al. [291] presented a data structure called *hierarchical Bloom filter* to support substring matching. This structures supports the checking of a part of string for containment in the filter with low false positive rates. The filter works by splitting an input string into a number of fixed-size blocks. These blocks are then inserted into a standard Bloom filter. By using the Bloom filter, it is possible to check for substrings with a block-size granularity. This substring matching may result in combinations of strings that are incorrectly reported as being in the set (false positives). For example, a concatenation of two blocks from different strings would be incorrectly recognized as an inserted substring.

Algorithm 7.3 Pseudocode for counting Bloom filter insertion

Data: x is the object key to insert into the counting Bloom filter.
Function: *insert(x)*
for $j : 1 \ldots k$ do
    /* Loop all hash functions k */

    $i \leftarrow h_j(x)$
    /* Increment counter C_i */

    $C_i \leftarrow C_i + 1$
    if $B_i == 0$ then
        /* Bloom filter had zero bit at position i */
        $B_i \leftarrow 1$
    end
end

Algorithm 7.4 Pseudocode for counting Bloom filter deletion

Data: x is the object key to be removed from the counting Bloom filter.
Function: *delete(x)*
for $j : 1 \ldots k$ do
   /* Loop all hash functions k */

   $i \leftarrow h_j(x)$
   /* Decrement counter C_i */

   $C_i \leftarrow C_i - 1$
   if $C_i \leq 0$ then
      /* Reset bit at position i */
      $B_i \leftarrow 0$
   end
end

The hierarchical Bloom filter construction improves matching accuracy by inserting the concatenation of blocks into the filter in addition to inserting them separately. This means that two subsequent single-block matches can be verified by looking up their concatenation. This approach generalizes to a sequence of blocks; however, storage space requirements grow as more block sequences are added to the structure.

This filter was used to implement a payload attribution system that associates excerpts of packet payloads to their source and destination hosts. The filter was used to create compact digests of payloads. The system works by dividing the payload of each packet into a set of blocks of a certain fixed size. Each block is appended by its offset in the payload: *(content||offset)*. The blocks are then hashed and inserted into a Bloom filter. A hierarchical Bloom filter is a collection of the standard Bloom filters for increasing block sizes.

When a string is inserted, it is first broken into blocks that are inserted into the filter hierarchy starting from the lowest level. For the second level, two subsequent blocks are concatenated and inserted into the second level. This block-based concatenation continues for the remaining levels of the hierarchy. The resulting structure can then be used to verify whether or not a given string occurs in the payload. The search starts at the first level and then continues upward in the hierarchy to verify whether the substrings occurred together in the same or different packets.

7.2.7 Spectral Bloom Filters

Spectral Bloom filters generalize Bloom filters to storing an approximate multiset. The membership query is generalized to a query on the multiplicity of an element. The answer to any multiplicity query is never smaller than the true multiplicity and is greater only with probability ϵ. The space usage is similar to that of a Bloom filter for a set of the same size (adding multiplicities).

Spectral Bloom filters extend Bloom filters to store an approximate multiset, and they support frequency queries [86]. The answer to a multiplicity query is never smaller than the actual value and is greater only with probability ϵ. The query time is $\Theta(\log(\frac{1}{\epsilon}))$. Spectral Bloom filters are used in storing shortest path distance information.

7.2.8 Bloomier Filters

Bloom filters have been generalized to *Bloomier* filters that compactly store functions. The Bloomier filters can encode functions instead of sets and allow the association of values with a subset of the domain elements [71]. Bloomier filters are implemented using a cascade of Bloom filters.

In more detail, given $S \subseteq D$, $n = |S|$ and a function $f : S \mapsto \{0, 1\}^k$, a Bloomier filter is a data structure that supports queries to the function value. It also has one-sided error: given $x \in S$, it always outputs the correct value $f(x)$ and if $x \subseteq D\backslash S$ with high probability it outputs $\perp$, a symbol not in the range of f [71]. In other words, the filter returns the desired function value for any element in the set S and returns the undefined value for any element not in S. For any elements not in S, there is a possibility of a false positive, in which case the filter may return an incorrect function value.

The query time of a Bloomier filter is constant and space requirement is linear. The basic construction of a Bloomier filter requires $O(n \log n)$ time to create, $O(n)$ space to store and, $O(1)$ time to evaluate. Although a Bloomier filter can be made mutable, the set S is immutable. This means that in a mutable Bloomier filter, function values can be changed but set membership (in S) cannot change.

7.2.9 Approximate State Machines

Efficient and compact state representation is needed in routers and other network devices in which the number and behavior of flows needs to be tracked. The *approximate concurrent state machine (ACSM)* approach was motivated by the observation that network devices, such as network address translation devices (NATs), firewalls, and application-level gateways, keep more and more state regarding TCP connections [37]. The ACSM construction was proposed to track the simultaneous state of a large number of entities within a state machine. ACSMs can return false positives, false negatives, and "do not know" answers. Their construction is based on Bloom filters and hashing.

7.2.10 Perfect Hashing Scheme

A simple technique called perfect hashing (or explicit hashing) can be used to store a static set S of values in an optimal manner using a perfect hash function. An array of size n stores the perfect hash value for each $x \in S$ and the information associated with x [249]. It follows from the definition of 2-universal hashing that any element y not in S has probability at most ϵ of having the same hash function value $h(y)$ as the element in S that maps to the same entry of the array.

A minimal perfect hash function is stored for S using $O(n + \log w)$ bits and a function $h : \{0, 1\}^w \mapsto \{0, 1\}^{\log(2/\epsilon)}$ from a 2-universal family using $O(\log n + \log w)$ bits. This results in $O(n + \log w)$ space and constant-time lookups. An array of size n is stored, where the entry that is the perfect hash value of $x \in S$ contains

1. the value $h(x)$ and
2. the information associated with x.

Figure 7.6 illustrates the perfect hashing technique. Lookup of x simply consists of computing a value of the perfect hash function and checking whether the stored hash function value is $h(x)$. This approach is not preferred for dynamic environments because the perfect hash function needs to be recomputed when the set S changes.

7.2.11 Summary

Figure 7.7 presents a comparison of the different Bloom filter variants discussed in this chapter. Bloom filters come in many shapes and forms, and they are widely used in distributed

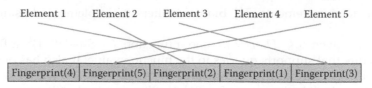

FIGURE 7.6
Example of explicit hashing.

systems due to their compact nature and configurable trade-off between size and accuracy. The basic Bloom filter offers a probabilistic representation of a set, but it does not support counting, deletion of elements, and multisets. The basic structure also has a number of other limitations. The lookup time grows as the false positive rate decreases (number of hash functions). The space usage of a Bloom filter is a factor $\log e \approx 1.44$ from the information theoretically optimal. Different variants have been developed to address these shortcomings.

The compressed Bloom filter improves on the basic construction by making it more friendly toward compression. The main expected usage for compressed Bloom filters is in transferring them over the network, and it requires an additional compression step—for example, using arithmetic coding.

The counting Bloom filter adds a counter to each bit in the filter, thus making it possible to count elements and remove them from the filter. The interesting result regarding counting filters is that a relatively small counter of 4 bits suffices for most requirements. This structure can also answer to multiplicity queries based on the counters. The d-left counting Bloom filter improves the basic construction by making it more balanced and thus results in better false positive rates.

Hierarchical Bloom filter uses basic Bloom filters in an hierarchical manner to be able to answer substring queries. As such, it does not support counting, deletion, and multisets. The

Filter	Description	Counting	Deletion	Multisets
Bloom filter (BF)	Probabilistic representation of a set	No	No	No
Compressed Bloom filter	Sparser BF compressed for transmission	No	No	No
Counting Bloom filter or d-left hashing with fingerprints	BF with c-bit counters	Yes	Yes	No
Hierarchical Bloom filter	Suitable for substring data	No	No	No
Spectral Bloom filter	BF that supports multiple sets, minimum counter	Yes	Yes	Yes
Bloomier filter	Probabilistic representation of multisets	Yes	No	Yes (functions), mutable function values
Perfect Hashing	Hashing technique that achieves near-optimal storage	No	No	No

FIGURE 7.7
Comparison of Bloom filters.

Spectral Bloom filter is a more advanced structure that supports both deletion, multiplicity queries (frequencies), and multisets. The Bloomier filter allows the insertion of functions to the structure, and it is based on a recursive cascade of Bloom filters. This structure can offer counting and multisets but does not allow the set of elements to change.

Finally, perfect (or explicit) hashing is a simple technique that achieves near-optimal storage, but it is suitable only for static environments and does not support deletion or counting.

7.3 Bloom Filters in Distributed Computing

We have surveyed techniques for probabilistic representation of sets and functions. The applications of these structures are manyfold, and they are widely used in various networking systems such as Web proxies and caches, database servers, and routers. In this section, we consider four types of network-related applications areas for these structures:

- Caching for Web servers and storage servers.
- Supporting P2P networks: Probabilistic structures can be used for summarizing content and caching [128].
- Packet routing and forwarding. Probabilistic techniques can be used to improve efficiency and scalability of various network processes.
- Supporting monitoring and measurement activities. Probabilistic techniques can be used to store and process measurement data summaries in routers and other network entities.

7.3.1 Caching

Bloom filters have been applied extensively to caching in distributed environments. To take an early example, Fan, Cao, Almeida, and Broder proposed the summary cache [127, 128] system, which uses Bloom filters for the distribution of Web cache information. The system consists of cooperative proxies that store and exchange summary cache data structures, essentially Bloom filters. When a local cache miss happens, the proxy in question will try to find out if another proxy has a copy of the Web resource using the summary cache. If another proxy has a copy, then the request is forwarded there.

In order for distributed proxy-based caching to work well, the proxies need to have a way to compactly summarize available content. In the summary cache system, proxies periodically transfer the Bloom filters that represent the cache contents (URL lists).

Dynamic content poses a challenge for caching content and keeping the summary indexes updated. Within a single proxy, a Bloom filter representing the local content cache needs to be recreated when the content changes. This can be seen to be inefficient; as a solution, the summary cache uses counting Bloom filters for the maintenance of their local cache contents and then, based on the updates, a regular Bloom filter is broadcast to other proxies.

The summary cache–based technique is used in the popular Squid Web Proxy Cache.[1] Squid uses Bloom filters for so-called cache digests. The system uses a 128-bit MD5 hash of the key, a combination of the URL and the HTTP method, and splits the hash into four equal chunks. Each chunk modulo the digest size is used as the value for one of the Bloom filter hash functions. Squid does not support deletions from the digest and thus the digest must be periodically rebuilt to remove stale information.

---

[1] www.squid-cache.org

Bloom filters have been applied extensively in distributed storage to minimize disk lookups. As an example, we consider Google's Bigtable system, which is used by many massively popular Google services (such as Google Maps and Google Earth) and Web indexing. Bigtable is a distributed storage system for structured data that has been designed with high scalability requirements in mind—for example, capability to store and query petabytes of data across thousands of commodity servers [68].

A Bigtable is a sparse multidimensional sorted map. The map is indexed by a row key, a column key, and a timestamp. Each value in the map is an uninterpreted array of bytes. Bigtable uses Bloom filters to reduce the disk lookups for nonexistent rows or columns [68]. As a result, the query performance of the database has to rely less on costly disk operations and thus performance increases.

7.3.2   P2P Networks

The exchange of keyword lists and other metadata between peers is crucial for P2P networks. Ideally, the state should be such that it allows for accurate matching of queries and takes sublinear space (or near-constant space). As discussed in Chapter 4, the later versions of the Gnutella protocol use Bloom filters [32] to represent the keyword lists in an efficient manner. In Gnutella, each leaf node sends its keyword Bloom filter to an ultra node, which can then produce a summary of all the filters from its leaves and then send this to its neighboring ultra nodes.

Bloom filters can be applied for approximate set reconciliation. This application is important for P2P systems, in which a peer may send a compact data structure to another peer that represents items that the peer already has. Bloom filters are not directly ideal for this kind of set-reconciliation application because of the possibility for false positives. Therefore, a number of Bloom filter–based structures have been developed [44, 270].

Rhea et al. [191] designed a probabilistic routing algorithm for P2P location mechanisms in the OceanStore project. Their aim was to determine when a requested file has been replicated near the requesting system. This system uses a construction called *attenuated Bloom filter*, which is an array of d basic Bloom filters. The ith basic filter keeps a record of what files are reachable within i hops in the network. The attenuated Bloom filter only finds files within d hops, but the returned paths are likely to be the shortest paths to the replica. In the distributed system, a node maintains attenuated filters for each neighbor separately, and updates are broadcast periodically.

The OceanStore system uses a two-tiered model in which the attenuated filter is part of the first tier. If the probabilistic search fails, the search can then fall back to a deterministic overlay search using Tapestry.

7.3.3   Packet Routing and Forwarding

Bloom filters can be applied in various parts in a routing and forwarding engine. Probabilistic techniques have been used for efficient IP lookups. IP routers forward packets based on their address prefixes. Each prefix is associated with the next hop destination. CIDR-based routing and forwarding uses longest prefix match for finding the next hop destination. This is commonly solved using a binary search, a trie search, or a TCAM. IP lookups can be made more efficient by dividing the addresses into tables based on their length and then utilizing binary search to find the longest common prefix. The d-left hashing technique has been used to make this lookup more compact and efficient [41].

Many different probabilistic structures have been developed for fast packet forwarding. To take one example, an algorithm that uses Bloom filters for *longest prefix matching* (*LPM*) was introduced in [110]. The algorithm performs parallel queries on Bloom filters to determine address prefix membership in sets of prefixes sorted by prefix length. This work indicates that Bloom filter–based forwarding engines can offer favorable performance characteristics compared to TCAMs used by many routers.

Bloom filters can be used for loop detection in network protocols. IP uses the time-to-live (TTL) field to detect and drop packets that are in a forwarding loop. The TTL counter is incremented for each network hop. For small loops, TTL may still allow a substantial amount of looping traffic to be generated.

Icarus is a system that uses Bloom filters for preventing unicast loops and multicast implosions. The idea is straightforward—namely, to use a Bloom filter in the packet header as a probabilistic loop detection mechanism. Each node has a corresponding mask that can be ORed with the Bloom filter in the header of a packet and then determine whether or not a loop has occurred. Detection accuracy can be traded off against space required in the packet header [347].

Bloom filters can also be used in multicast forwarding engines. A multicast packet is sent through a multicast tree. A multicast router maps an incoming multicast packet to outgoing interfaces based on the multicast address. Initially, Grönvall suggests an alternative multicast forwarding technique using Bloom filters [151]. In this technique, a router has a Bloom filter for each outgoing interface. The filters contain the addresses associated with the interfaces. When a multicast packet arrives on one interface, the Bloom filters of each interface are checked for matches. The packet is forwarded to all matching interfaces. This technique is interesting because it does not store any addresses at the router; however, the addition or removal of multicast addresses requires that the Bloom filters be updated. A similar technique has been proposed recently for publish/subscribe networking [121].

Bloom filters have also found applications in deep packet scanning, in which applications need to search for predefined patterns in packets at high speeds. Bloom filters can be used to detect predefined signatures in packet payload. When a suspect packet is encountered, it can then be moved for further investigation. One advantage of Bloom filters is that they can be efficiently implemented in hardware and parallelized [109].

7.3.4 Measurement

Bloom filters have found many applications in measurement of network traffic. One particular application is the detection of heavy flows in a router. Heavy flows can be detected with a relatively small amount of space and small number of operations per packet by hashing incoming packets into a variant of the counting Bloom filter and incrementing the counter at each set bit with the size of the packet. Then, if the minimum counter exceeds some threshold value, the flow is marked as a heavy flow [129].

Iceberg queries have been an active area of research development. An iceberg query identifies all items with frequency above some given threshold. Bloom filter variants that are able to count elements are good candidate structures for supporting iceberg queries. In networking, low-memory approximate histogram structures are needed for collecting network statistics at runtime. For example, in some applications it is necessary to track flows across domains and perform, to name a few examples, congestion and security monitoring. Iceberg queries can be used to detect denial-of-service attacks.

Packet and payload attribution is another application area in measurement for Bloom filters. The problem in payload attribution is as follows. Given a payload, the system reduces the uncertainty that we have about the actual source and destination(s) of the payload within a given target time interval. The goodness of the system is directly related to how much this uncertainty can be reduced. The implementation of a payload attribution system has two key components—a payload-processing component and a query-processing component.

The *source path isolation engine* (*SPIE*) [300] implements a packet attribution system in which the system keeps track of incoming and outgoing packets at a router. Simply storing all the resulting information is not feasible. Therefore, Snoeren et al. proposed using Bloom filters to reduce the state requirements. A Bloom filter stores a summary of packet information in a probabilistic way. One key observation is that each router maintains its own Bloom filters and thus their hash functions are independent.

A SPIE-capable router creates a packet digest for every packet it processes. The digest is based on the packet's nonmutable header fields and a prefix of the first 8 bytes of the payload. These digests are then maintained by a network component for a predefined time.

When a security component, such as an intrusion-detection system, detects that the network is under attack, it can use SPIE to trace the packet's route through the network to the sender. A single packet can be traced to its source, given that the routers on the route still have the packet digest available. A false positive in this setting means that a packet is incorrectly reported as having been seen by a router. When the source of a packet is traced, false positives mean that the reverse path becomes a tree (essentially branches to multiple points due to false positives).

The packet attribution was extended to payload attribution by Shanmugasundaram et al. [291] with the hierarchical Bloom filter. As discussed earlier in this chapter, this structure allows the query of a part of a string. SPIE uses the nonmutable headers and a prefix of the payload, whereas with hierarchical Bloom filters it is sufficient to have only the payload to perform a traceback.

7.4   Gossip Algorithms

Probabilistic *gossip protocols* have gained considerable interest in recent years, starting with the seminal work of Alan Demers et al. in 1987 [108]. A gossip protocol is based on opportunistic interactions between nodes, in a manner similar to gossip in social networks or the way in which a viral infection spreads in a biological population [184]. The mathematics of epidemics are frequently used to model gossip-based systems, and the term *epidemic algorithm* has been used to describe gossip-like software systems. They can be applied in various distributed environments where nodes are expected to communicate frequently with other nodes. The probabilistic nature of gossip allows the flooding of a message with a relatively low cost [30].

> Epidemic and gossip algorithms have been recently recognized as robust and scalable means to disseminate information in wide-area environments. Information is disseminated reliably in a distributed system the same way an epidemic would be propagated throughout a group of individuals. Each process of the system chooses random peers to which information is forwarded. The underlying P2P communication paradigm is the key to the scalability of the dissemination schemes.

7.4.1   Overview

Gossip-based unstructured overlays can be used as a building block for various network and service management applications, especially when support for dynamic operation and eventual consistency is required [335]. Gossip has been used for monitoring and configuration in the AstroLabe system [331] and for achieving eventual consistency in Amazon's Dynamo [105], which is a core system for the company's services. Gossip lends itself well to monitoring in large-scale networks where each node monitors a small random subset of other nodes, thus distributing the monitoring cost. Gossip is also suitable for managing routing tables in a large-scale P2P network.

Gossip-based protocols can be divided into three main categories:

- *Dissemination* protocols, which use gossip to spread data across the network by using probabilistic flooding. A technique called *rumormongering* uses gossip for some predetermined time that has been chosen high enough to ensure that the gossiped information is sent across the network to all expected receivers.
- *Anti-entropy* protocols, which are used to replicate data. These protocols compare replicas and reconcile differences in an opportunistic fashion. An anti-entropy protocol gossips information until it is made obsolete by newer information.
- *Data aggregation* protocols, which compute a network-wide aggregate by sampling information at the nodes. The aim is to ultimately compute a system-wide aggregate value, for example, the largest measurement value. In order to efficiently aggregate computation to work, the aggregate function must be computable by fixed-size pairwise information exchanges. Typically the exchanges eventually terminate after a logarithmic number of rounds to the system size. Aggregation-based protocols can be used also to implement sorting, counting, and summing of values at nodes [171].

The first category includes various kinds of multicast and event-dissemination protocols. A gossip dissemination can be triggered periodically (push or pull) or when an external request is received. The key concern is the latency of the communication and the probability of reaching all proper nodes in the network through gossiping. Latency is a concern especially with periodic operation. The random choice of the subset of nodes to contact can be determined using local information acquired by a node during its execution. Gossip protocols that utilize local information in peer selection are called *informed gossip protocols*.

Figure 7.8 presents an overview of an abstract model for gossiping. The gossip is either triggered by an application reacting to some event (1) or by a periodic trigger. The gossip is driven by a predefined policy that needs to first decide whether or not the gossip process is started. The gossip engine consults this policy and makes a gossip decision (2). The gossip decision is based on the current state of the system and the policies.

After the decision has been made, the gossip engine needs to determine where gossip messages are sent. This involves peer selection. Different policies can be used for selecting peers. A subset of peers can be selected randomly or based on an estimate of their reliability and lifetime on the network. As discussed in Chapter 4, peer selection is an integral part

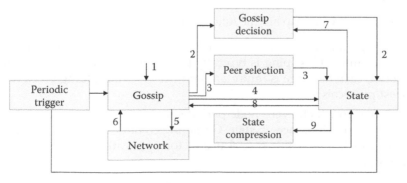

FIGURE 7.8
Key gossip interactions.

of unstructured P2P protocols. Peer selection also consults state (3), and a list of candidate peers is then created as a result (4). Many gossip algorithms may need to decide the content to be gossiped. For example, a proactive gossip protocol retrieves the parameters from its current state.

The gossip engine then uses the underlying network to communicate with the selected peers. The communication can utilize unicast, multicast, or broadcast (5). The network informs the engine about message delivery (6). If the engine is informed that certain messages could not be delivered and there are communication problems, it can then consult the peer selection and find other peers to contact. The interactions between the gossip engine and the network depend on the operating environment. For Internet applications, the gossip engine can utilize a bootstrapping server to find peers. In pervasive and ubiquitous computing environments, the gossip engine may rely on peer advertisements from the network (such as *universal plug and play (UPnP)*) discovery messages.

When a gossip message is received from the network, a gossip protocol may react in different ways to the message. The received message may be forwarded to the current peers. The message may result in a response—for example, sending the requested information pertaining to the current state. State contained in the message may be extracted and combined or compared with the current state (7). The gossip protocol can then decide whether to merge the received state with its local state or not. If merging takes place, the new state can then be either proactively or reactively disseminated (8). State can also be compressed (9) to improve both local and distributed processing [2].

7.4.2 Design Considerations

The gossip process forwards a message each time to a randomly selected set of peers. A crucial observation for the reliability of the system is that the peer sets of nodes in the system are independent. This replication factor of a message is called the *fanout* of the dissemination, and it is a key parameter for gossip and epidemic algorithms. The reliability of these algorithms is based on their proactive nature, in which redundancy and randomization circumvent potential failures and disruptions in communications. Gossip and epidemic algorithms avoid expensive reconfiguration when failures happen because they rely on the probabilistic dissemination and the fact that the message is eventually delivered across the network to the proper receivers.

Epidemic algorithms exhibit a bimodal behavior. There is a threshold in the parameter configuration below which a reliable delivery can be ensured. The delivery guarantee depends on the system parameters.

7.4.3 Basic Models

The two basic models of gossip are the following:

- Rumormongering, in which nodes periodically choose a node at random and spread the rumor. Gossiping is typically performed for some preset duration. In this model, each message is important and thus it is a building block for reliable multicast protocols.

- Anti-entropy, in which every node periodically chooses another node at random and resolves any differences in state. Gossiping is continued until the data becomes obsolete, which makes this model useful for applications that require eventual consistency.

Anti-entropy supports the replication of state that does not have strong consistency requirements. Updates are distributed by the participating nodes, and the expected

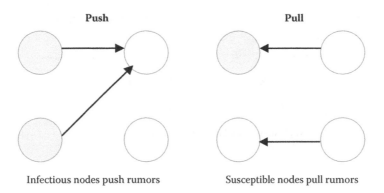

FIGURE 7.9
Examples of push and pull interactions.

dissemination time grows logarithmic to the number of nodes in the system even when failures occur. Anti-entropy protocols gossip information until it is made obsolete by newer information. Rumormongering uses gossip for some predetermined time that has been chosen high enough to ensure that the gossiped information is sent across the network to all expected receivers.

In rumormongering, nodes have three states, infected, susceptible, and removed. In the first state, a node has a certain piece of data and gossips with other nodes to propagate this data. With the second state, a node does not yet have the data but is willing to gossip. In the third state, a node that has already received the data has been removed with some probability. A removed node does not participate in gossiping. A node running the rumor-mongering algorithm simply picks another node at random and exchanges updates with this node.

There are two central types of interactions for gossip, namely push and pull. Figure 7.9 illustrates these two modes. In push, nodes can directly contact other nodes, their peers, and send information. In pull-based gossip, a node sends a digest (summary) of its state to another node and requests updates. The other node then checks what information should be sent to the requesting node and sends the appropriate content. The two modes can be combined for push-pull gossip, in which the responder includes a list of information elements that appear to be outdated based on the digest. The requestor can then send the updates to this node.

Pushing involves one message, pulling involves two messages, and the push-pull cycle involves three messages. Pushing is not a very good choice when rapid dissemination is required. The pull mode works better when many nodes are infected (have the data). The hybrid mode is the most efficient of the three modes since it allows information to propagate faster. Gossiping is divided into rounds. Gossiping takes $O(\log n)$ rounds to propagate a single update to all nodes.

Gossip systems can be extended to support the deletion of data. This is achieved by creating a special record for deleted data called a *death certificate*. The death certificates are timestamped and distributed in the environment to inform nodes that a certain piece of data has been removed. In order to keep the data from being reinjected into the system, some nodes need to keep a permanent copy of the death certificate.

7.4.4 Basic Shuffling

Figure 7.10 illustrates the general view-based gossiping model, in which nodes have partial views regarding the P2P network and randomly exchange views. Basic shuffling is a simple

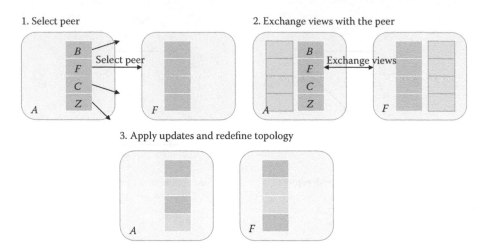

FIGURE 7.10
Example of view shuffling.

gossip algorithm that uses a push-pull strategy [306]. The algorithm assumes that there are no failures and that neighborhood information is available [19]. The idea is to form an overlay and keep it connected by means of an epidemic algorithm. In the protocol, each peer keeps a set of continuously changing neighbors and occasionally contacts a random neighbor to exchange some of their neighbors. An entry in the neighbor table contains the network address (IP address and port) of another peer in the overlay. Each peer periodically initiates the *shuffle* operation with a neighbor.

The shuffle operation has the following six steps from the viewpoint of the initiating node *A*:

1. Select a random subset of *t* neighbors from the local neighbor list. Select a random peer, *B*, within the subset. The parameter *t* is called the *shuffle length*.
2. Replace *B*'s address with *A*'s address.
3. Send the updated subset to *B*.
4. Receive from *B* a subset of no more than *t* of *B*'s neighbors.
5. Discard entries pointing to *A* and entries that are already in *A*'s neighbor table.
6. Update *A*'s neighbor table to include all remaining entries. First populate empty slots and then replace entries that were originally sent to *B*.

When a shuffling request is received by a peer, it randomly selects a subset of its own neighbors and sends it to the initiating node. Then the node executes steps 5 and 6 to update its neighbor table.

The shuffling operation reverses the relation between *A* and *B*. After node *A* has initiated a shuffling operation with its neighbor *B*, *A* becomes *B*'s neighbor, while *B* is no longer a neighbor of *A*.

Given a fail-free environment, the connectivity resulting from the shuffle algorithm is guaranteed. A node cannot become disconnected because of the shuffling operation, because it only changes the neighbors. An overlay built using shuffling cannot be split into two disjoint subsets as a result of the shuffling operation [335].

7.4.5 Enhanced Shuffling

The Cyclon overlay system introduced an enhanced version of shuffling that improves the quality of the algorithm in terms of randomness [335]. The key difference between basic shuffling and enhanced shuffling is that, in the latter, nodes do not randomly choose which neighbor to shuffle with; instead they select the neighbor that has the most recent information.

Enhanced shuffling nodes initiate neighbor exchanges periodically. In enhanced shuffling, the neighbor table elements contain an extra field called age that denotes the age of the entry expressed in intervals since its creation by the node it references. The enhanced shuffling operation involves the following seven steps from the viewpoint of the initiator A:

1. Increase the age of all neighbors by one.
2. Select neighbor B with the highest age among all neighbors and a random subset of neighbors of size $t - 1$.
3. Replace B's entry with a new entry of zero age and with A's address.
4. Send the updated subset to peer B.
5. Receive from B a subset of at most t entries.
6. Discard entries pointing at A and entries already contained in A's neighbor table.
7. Update A's neighbor table to include all remaining entries. First populate empty slots and then replace entries that were originally sent to B.

As in the case of basic shuffling, when a shuffling request is received by a peer, it randomly selects a subset of its own neighbors, of size at most t, and sends it to the initiating node. Then the node executes steps 5 and 6 to update its neighbor table. The age count is not increased by the receiver; it is only increased by the initiator.

The Cyclon protocol results in a graph that has similar properties to those of random graphs in terms of the average path length, clustering coefficient, and diameter. Random peer sampling can be replaced with a sampling algorithm that maintains shortcuts to far-away peers. Random shortcut selection with greedy routing results in $n^{1/3}$ average hop count, where n is the number of peers in the small-world topology. Using Kleinberg' small-world routing results in greedy routing performance of $O(\log^2(n))$. It is possible to instrument the peer sampling to achieve routing similar in performance to Kleinberg's greedy routing [183]. Experimental results with P2P gossip-based protocols indicate that small-world topologies with randomly chosen shortcuts perform reasonably well in practice [36].

7.4.6 Flow Control and Fairness

As in the case of network protocols in general, gossip algorithms need to have a mechanism to ensure fairness. The goal of a flow control mechanism for gossip is to determine in an adaptive fashion the maximum rate at which a participant can send updates without creating a backlog of updates [332]. A flow control mechanism should be fair and allow each participant to send updates even when the system is under heavy load. Since there is no global control, the flow control system needs to be decentralized.

Flow control can be accomplished through the gossip protocol by having each participant maintain a maximum update rate. When two participants gossip, they exchange the maximum update rates and split the difference between the maximum rates. A similar technique that is used in TCP (additive increase, multiplicative decrease) can be used when a message overflows or underflows. If a gossip message overflows, then the maximum rate is reduced by a percentage. If a gossip message underflows, then the rate can be additively increased [332].

7.4.7 Gossip for Structured Overlays

Gossip protocols are a form of unstructured overlays. Gossip can be used together with a structured overlay to improve system scalability. Gossiping can offer eventual consistency for a wide-area system. It has been shown that the randomness provided by an unstructured gossip overlay can be used to build the routing table of a structured P2P overlay. This system uses the leaf set of Pastry and the proximity links of an unstructured overlay to build a complete overlay. Simulation results of the system indicate that the combination of the two overlay techniques can be used to significantly reduce overlay maintenance overhead without adverse effects to performance [218].

Gossip has been applied for multicast and pub/sub systems, in which the goal of the system is to deliver messages from publishers to the subscribers. The motivation for gossiping in pub/sub is that gossip protocols are simple and do not require a pub/sub routing infrastructure. The limitations include more overhead in communications. A number of gossip algorithms for pub/sub systems have been proposed in [18, 22, 74, 92, 124, 253, 336].

Eugster and Guerraoui present the *probabilistic multicast (pmcast)* system that is an example of informed gossip for pub/sub event delivery [123]. The system avoids gossiping to subscribers who are not interested in the content. This is achieved by organizing nodes in a hierarchy of groups, which are built based on the physical proximity of nodes.

Event messages are disseminated using the hierarchy by gossiping depth-wise, starting at the root. The hierarchy is mapped to network topology, which allows for reduction of the number of network boundaries that are crossed during the multicast. Each node maintains a view that includes the subscriptions of its neighbors in a group. Special members of the group called delegates are responsible for aggregating subscriptions within a group and have access to the other views of other nodes at the same level of hierarchy. Membership information updating is based on gossip pull.

Voulgaris et al. present a multilayer architecture called *SUB-2-SUB* where content-based event dissemination is realized by traversing multiple layers [336]. This architecture is based on three layers. The lower layer uses a gossip protocol to exchange information pertaining to subscriptions. The middle layer is responsible for maintaining semantic relations between the subscribers and clustering them based on their interests. The upper layer is a logical ring structure that connects all participants.

The system leverages the overlapping intervals of range subscriptions and creates an unstructured overlay reflecting the structure of the attribute space and that of the set of subscriptions. Once subscriptions are clustered, events are directly posted to the proper cluster where they are delivered. The SUB-2-SUB system uses Cyclon for gossiping. An event publisher needs to find at least one subscriber for the event being published. After this, the subscribers collaborate in the dissemination by using the shortcuts links obtained using gossiping and the ring topology. The ring ensures that all potential subscribers are reached.

8

Content-based Networking and Publish/Subscribe

Content-based routing has become an active research area. In this chapter we consider content-centric routing and examine a number of protocols and algorithms. Special emphasis is placed on distributed publish/subscribe (pub/sub), in which content is targeted to active subscribers. A content-based router is part of an overlay structure in which each router forwards events to neighboring routers and local clients based on their interests.

In this chapter, we give an overview of content-based pub/sub systems and focus on content-centric routing and forwarding operations performed by a router. First we give an overview of different data-centric and content-based systems that have been developed. Then we focus on the Siena and Hermes systems, where the former is based on a static router topology and the latter uses a distributed hash table (DHT) to be able to support more dynamic environments. The Siena content-based router is used as an example, and a number of optimization techniques for improving performance are discussed. The chapter also discusses the formal specification of pub/sub systems and how mobile subscribers and publishers can be supported.

8.1   Overview

Figure 8.1 presents the key data-centric and content-based systems discussed in this chapter. They have been positioned based on the expressiveness on the x axis and the dynamics of the supported topology on the y axis. The middle row of the diagram consists of DHT-based solutions—Scribe, Bayeux, and i3 are examples of DHT-based data-centric solutions. DONA is a data-centric anycast system that follows the current Internet structure by introducing hierarchical anycast request-processing routers into the network. It therefore provides more rigid structure to the network. SplitStream is an example of a DHT-based system for efficient multicast, and it explicitly addresses the concerns with dynamic operation and offers redundancy. In addition, a number of structured gossip-based overlays have been proposed. Gossip-based techniques were discussed in Section 7.4.

Both Meghdoot and MEDYM are more content-based, and the former builds on a DHT whereas the latter uses a novel clustering technique that can use various underlying multicast solutions. On the right side, we have proper content-based systems that allow more complex queries and query aggregations. Siena is a classic example of content-based routing based on a static topology of application-layer routers. Hermes is a DHT-based system that follows the Siena model and uses rendezvous points to coordinate both signaling and message propagation. Hermes therefore can support more dynamic environments. Finally, we have content-based systems designed for dynamic environments that allow reconfiguration between the routers. The scalability to Internet-wide environments of these different proposals is an open issue. DHT-based solutions can be seen as possible candidates due to the flexibility and scalability offered by the basic overlay substrate.

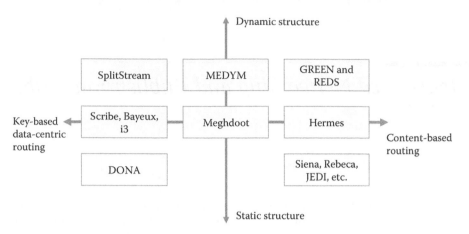

FIGURE 8.1
Overview of data-centric and content-based systems.

8.2   DHT-based Data-centric Communications

In this section, we examine a number of DHT-based data-centric systems, with the focus on efficient wide-area anycast and multicast. Many of the systems are based on rendezvous-based routing [344] such as Scribe [66], Bayeux [365], SplitStream [64], OverCast [170], Meghdoot [155], MEDYM [155], and i3 [307]. These systems are motivated by the observation that introducing some control in the form of fixed processing points results in improved multicast tree management operations.

8.2.1   Scribe

Scribe is a scalable application-level multicast infrastructure built on top of the Pastry DHT [66]. Any Scribe node may create a group, and other nodes can then join the group. Nodes that are members of the group can then multicast messages to all members of the group. Scribe provides best-effort delivery of multicast messages and does not enforce any particular delivery order for the messages.

Each Scribe multicast group is represented by a Pastry key called the *groupId*. A multicast tree for a given groupId is created by taking the union of the Pastry routes from each group member to the groupId's root. Content can then be sent using this multicast tree by using reverse path routing from the root toward the leaves of the tree.

Scribe relies on the properties of the underlying Pastry substrate for efficiency. The delay to forward a message from the root to each group member is low because of the low delay penalty of Pastry routes. The local route convergence property of Pastry ensures that the load imposed on the physical network is small. This is because most message replication is performed by intermediate nodes that are close to the leaf nodes in the tree.

The group membership management is decentralized and efficient because it builds on the existing, proximity-aware Pastry overlay. The introduction of new members to the multicast tree is easy. A new member simply sends a message to the groupId. Thus Scribe can support large numbers of members per group. The groups can also be dynamic.

Pastry's proximity-aware routing and Scribe's multicast group management can be combined to support anycast communications. Anycast is useful when performing resource discovery. With Anycast, any node in the overlay can send an anycast message to a Scribe group. The anycast message is routed toward the groupId and forwarded to the nearest member by relying on the local route convergence property.

The Scribe multicast routing state is distributed and maintained in a decentralized fashion. Each node in a tree only maintains its immediate predecessors and successors in the tree. This can be seen as a significant scalability advantage over other overlay multicast schemes such a Bayeux [365]), discussed next. As a result, Scribe does not require excessive signaling traffic in order to gather global state information.

8.2.2 Bayeux

Bayeux is an application-level multicast protocol that has been built on top of the Tapestry DHT [365]. The Bayeux algorithm is similar to the Scribe algorithm because they both use an overlay routing layer to build a multicast tree for a given topic. The main difference from Scribe is in the way that the multicast tree is constructed. Bayeux sends a subscription message always to the root of the tree. The root node maintains the membership list for the topic-based multicast group. A response message from the root installs state in the intermediate overlay nodes that forward the message. Thus the new node becomes part of the forwarding tree.

Scribe can be seen to have a more scalable approach to the construction of the multicast tree. In Bayeux, the root node has to keep membership information pertaining to all members of the multicast group. Moreover, group membership management introduces overhead, since each control message must be sent to the root and the root then sends the reply back. To prevent the root node from becoming a bottleneck for performance, a partitioning scheme for the multicast tree has been proposed that shares the load among several root nodes.

8.2.3 SplitStream

SplitStream addresses the scalability of application-layer tree-based multicast [64]. The aim is to support efficient multicast when nodes participating in the peer-to-peer (P2P) network come and go. The solution is based on striping the content across a forest of *interior-node-disjoint* multicast trees. These trees distribute the forwarding load among participating peers [64]. SplitStream has been implemented on top of Scribe and Pastry; however, it could also be implemented using a different underlying overlay algorithm.

Many multicast systems are based on trees. In these systems, a node is either an interior node or a leaf node. Given a balanced tree with fanout f and height h, the number of interior nodes is given by $(f^h - 1)/(f - 1)$. The percentage of leaf nodes increases with f. The potential outbound bandwidth of the interior nodes is proportional to the fanout degree.

The key idea in SplitStream is to split the content into k stripes. A separate multicast tree is used to distribute a given stripe. Peers join as many multicast trees as there are stripes that they wish to receive. Each peer also defines an upper bound on the number of stripes they can forward to other peers. Given that the original content has bandwidth requirement B, each stripe has a bandwidth requirement of B/k. The peers can control their inbound bandwidth in increments of B/k.

Figure 8.2 illustrates SplitStream's forest construction. The source generates the stripes from the content and multicasts each stripe using its designated tree. The stripe identifier of each stripe starts with a different digit. The node identifiers of interior nodes share a prefix with the stripe identifier. This means that they must be leaves in the interior-node-disjoint multicast forest.

The challenge is to construct this forest of multicast trees such that an interior node in one tree is a leaf node in all the remaining trees and the bandwidth constraints specified by the nodes are satisfied. This ensures that the forwarding load can be spread across all

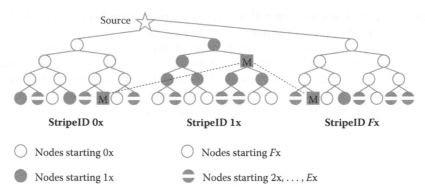

FIGURE 8.2
SplitStream forest construction.

participating peers. A set of trees is said to be interior-node-disjoint if each node is an interior node in, at most, one tree and a leaf node in the other trees. SplitStream exploits the properties of Pastry routing to construct interior-node-disjoint trees. k Scribe trees have a disjoint set of interior nodes when the identifier for the trees all differ in the most significant digit. The value of b for Pastry needs to be chosen so that it results in a suitable value for k.

For example, if all nodes wish to receive k stripes and they are willing to forward k stripes, SplitStream will construct a forest such that the forwarding load is evenly balanced across all nodes while achieving low delay and link stress across the system.

The following equation defines a rough upper bound on the probability of failure [64],

$$|N| \times k \times \left(1 - \frac{I_{min}}{k}\right)^{\frac{C}{k-1}}, \tag{8.1}$$

where N is the set of nodes, I_{min} is the minimum number of stripes that a node receives, and C is the total amount of spare capacity. The observation is that the probability of success is very high even with a small amount of space capacity.

The forest does not necessarily satisfy the constraints of nodes on outbound bandwidth. Therefore, SplitStream utilizes an algorithm to resolve the case where a node that has reached its outdegree limit receives a join request from a prospective child. Initially, the node accepts the prospective child. Then, it evaluates its new set of children to select a child to reject based on the stripe identifiers. If a reject node has not found a parent, it sends an anycast message to a special Scribe group called the spare capacity group. This group is used to find potential parents for orphan nodes.

In order to use the SplitStream system, applications need to encode content in such a way that each stripe requires approximately the same bandwidth and that the content can be reconstructed from any subset of the stripes of sufficient size. These requirements motivate the use of coding techniques in combination with the system—for example, erasure coding, which allows reconstruction of an original data block using sufficient subset of the block.

The expected amount of state maintained by each node is $O(log|N|)$. The expected number of messages to build the forest is $O(|N|log|N|)$, if the trees are well balanced, or $O(|N|^2)$ in the worst case. We expect the trees to be well balanced if each node forwards the stripe whose identifier shares the first digit with the nodes identifier to at least two other nodes.

8.2.4   Overcast

A similar system called Overcast provides scalable and reliable single-source multicast using a protocol for building data distribution trees that adapt to network conditions [170].

Overcast organizes dedicated servers into a source-rooted multicast tree using bandwidth-estimation measurements to optimize bandwidth usage across the tree. The main differences between Overcast and SplitStream are that Overcast uses dedicated servers while SplitStream utilizes clients. Moreover, SplitStream assumes that the network bandwidth available between peers is limited by their connections to their ISP rather than the network backbone.

8.2.5 Meghdoot

Meghdoot [155] is one of the early examples of a content-based pub/sub systems entirely based on a structured overlay infrastructure, namely content a addressable network (CAN) [267], with rendezvous-based event routing. Meghdoot uses structured subscriptions with either numerical or string attributes. A subscription is mapped to a CAN point whose coordinates are the bounds of each range constraints. A published event is mapped to a CAN region spanning all the possible subscriptions that can map to the event. A generic architecture for content-based pub/sub independent of the specific infrastructure has also been proposed [21].

8.2.6 MEDYM

Match early with dynamic multicast (MEDYM) [47] partitions the event space into nonoverlapping partitions with balanced load. Each server acts as a matcher for one or more partitions. A channelization technique is presented that partitions the event space into a number of multicast groups. A multicast tree is built for each group that spans servers with subscriptions for any event in that group. Multicast can be performed either through IP multicast, if available, or with application-level multicast.

8.2.7 Internet Indirection Infrastructure

The *internet indirection infrastructure (i3)* [307] is a Chord-based overlay network that aims to provide a more flexible communication model than the current IP addressing [307]. In i3, each packet is sent to an identifier. Packets are routed using the identifier to a single node in the overlay system. One of the i3 nodes is responsible for an address space in which the destination belongs. This node maintains triggers, which are installed by receivers that are associated with identifiers. When a matching trigger is found, the packet is forwarded to the corresponding receiver. The system is flexible in the sense that an i3 identifier can be used to represent various distributed entities such as hosts, objects, and sessions.

The i3 system can support a number of interactions, including unicast, multicast, anycast, and service composition. The overlay provides a level of indirection that can be used for supporting mobile and multihoming hosts. In i3 unicast, a host R inserts a trigger *(id, R)* in the i3 infrastructure to receive all packets with identifier *id*. In multicast, an application can build a multicast tree using a hierarchy of triggers. i3 provides support for anycast by allowing applications to specify a prefix for each trigger identifier. Packets are then matched to the identifiers according to the longest matching prefix rule.

8.2.8 Data-oriented Network Architecture

Data-oriented network architecture (DONA) [189] aims to introduce data-centric operations to the networking architecture. DONA introduces a data-handling shim layer above the network layer and support anycast queries by resolving them to nearest replicas. DONA does not use domain name system (DNS), but rather routes using the names given to data objects. The architecture introduces two new network entities: the data handlers that operate at the data-handling layer and perform name-based routing and caching, and the authoritative resolvers, which can point to the authoritative copy of a principal's data. The two new network primitives are *fetch(name)* and *register(name)* [189].

DONA's name-based anycast primitive can be used in many different types of resource discovery. For example, it can provide the basic primitives underlying session initiation protocol (SIP), can support host mobility and multihoming, and can establish forwarding state for interdomain multicast. The implementation of anycast at the naming layer rather than the network layer is motivated by separation of concerns. Name resolution is a control plane operation and thus it does not need to operate at link speeds.

8.2.9   Semantic Search

Adding support for range queries and semantic queries over DHTs has gained interest recently. This is essentially taking DHT technology toward content-based routing. In this section, we summarize two systems toward this end, namely the distributed segment tree and the pSearch system.

DHT-based systems have been enhanced to support more expressive search techniques. While these systems can offer search guarantees, they also require maintenance of their structured overlays. The search characteristics and performance of DHT-based structured keyword search systems are influenced by two factors:

- Indexing scheme that is used by the DHT. An index should be able to take data locality into account.
- Efficiency of the distributed query engine.

8.2.10   Distributed Segment Tree

A range query is intended to find all the keys in a certain range. A number of techniques for implementing range queries in P2P overlays have been developed. Mercury uses a circular overlay and stores data continuously in order to support multiattribute range query [28]. Skip graphs presented in Chapter 6 are distributed data structures that support range search. Space-filling curves and kd-trees also have been proposed for multidimensional queries [142]. In this section, we consider the distributed segment tree as an example technique. A dual query to a range query is the cover query, which aims to find all the ranges currently in the system that cover a given key.

The distributed segment tree is a layered DHT structure that incorporates the concept of segment tree. Figure 8.3 illustrates the segment tree structure. The binary segment tree uses

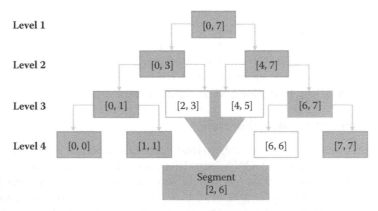

FIGURE 8.3
An example of a segment tree.

$O(n \log n)$ storage and can be built in $O(n \log n)$ time. The intervals that contain a given query point can be located in $O(\log n + k)$, where k is the number of retrieved segments. The height of the structure that supports range length L is $H = \log L + 1$. The structure also generalizes to higher dimensions. In the example, the range $[0, 7]$ has been divided into four levels. The range $[2, 6]$ is represented using ranges $[0, 7]$, $[0, 3]$, and $[4, 7]$.

The segment tree structure can then be distributed onto DHT by assigning the node interval $[s, t]$ to the DHT node with the key $h([s, t])$, where h is a hash function used to generate node identifiers. This means that information pertaining to any node of the segment tree can be efficiently located using a DHT lookup. Each node can reconstruct the range tree locally since the structure is a binary tree and can map each node in this local tree to distributed DHT nodes. This technique connects the structural information of the node intervals and the underlying DHT, in which structural information is not used in node identifiers. As a result, both range and cover query can be performed efficiently [286].

Insertion of a value requires the creation of a new leaf node in the distributed range tree (and mapped to a DHT node) and then propagating this information to all ancestors of the leaf node. The value is inserted to all covering nodes. Insertion of a cover query is similar, but differs in that it requires the insertion of a range. A range is inserted by decomposing it into corresponding segments, and then inserting these segments. A cover query does not require the ancestor propagation step that is needed when a value is inserted [286].

The system uses parallelism to achieve approximately $O(1)$ query complexity. Given a range to query, the range is first split into a union of minimum node intervals of segment tree using a specific range splitting algorithm. Then the identifiers for each of the ranges are obtained by using hashing. Each of the identifiers are then queried using a DHT lookup. The final answer is a union of the lookup answers.

8.2.11   Semantic Queries

The pSearch system is a decentralized structured P2P information retrieval system. pSearch uses a technique called *latent semantic indexing (LSI)* for generating descriptions of document semantics, and then distributing this information over the P2P network. The aim is to reduce the search cost of the system for a given query by taking document semantics into account [319].

Figure 8.4 gives an overview of the CAN-based pSearch system. The basic idea in pSearch is to define a semantic overlay and mapping the overlay nodes to physical nodes using the

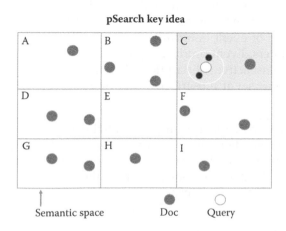

FIGURE 8.4
Overview of pSearch.

CAN DHT algorithm. Latent semantic indexing is used to position each document in a semantic Cartesian space. Documents that are close in the space have similar contents. Each query can also be positioned in this semantic space. A query involves comparing documents in a small region centered at the query. The dimensionality of CAN is set to be equal to that of LSI's semantic space. The semantic vector of a document is used as a key in CAN to store the document's index. This index includes the semantic vector and a reference to the document (URL).

The approach has to address the problem known as the *curse of dimensionality* that complicates the mapping of the high-dimensional space to practical dimensions that can be achieved with CAN. Moreover, the semantic vectors are not uniformly distributed in the semantic space. Thus direct mapping can result in unbalanced distribution. The pSearch system requires some global statistics such as the inverse document frequence and the basis of the semantic space. This information can be distributed in the network [4, 84, 95, 318].

The pSearch systems works as follows:

- When a new document is introduced to the system, a semantic vector is generated for the document using latent semantic indexing. This vector is used as a key to store the index in the CAN.

- When a query is received, a semantic vector is generated for the query, and it is routed in the overlay using the vector as the key.

- When the query reaches its destination defined by the key, the query is flooded to nodes within radius r. This radius is determined by the similarity threshold or the number of requested documents.

- Nodes that receive the query perform a local search using latent semantic indexing and report the matching references back to the user.

8.3   Content-based Routing

Event-based systems [45, 57, 118, 235, 244, 314, 364] are seen as good candidates for supporting distributed applications in dynamic and ubiquitous environments because they support decoupled and asynchronous one-to-many and many-to-many information dissemination [99, 259]. Event systems are widely used because asynchronous messaging provides a flexible alternative to remote procedure call (RPC) [87, 122].

In the general model of event notification, subscribers subscribe events by specifying their interests using filters. Event producers publish events (also known as notifications), which are matched against active subscriptions. Event *filtering* or *matching* is used to deliver information to the proper set of subscribers [7, 48, 54, 56, 59, 60, 125, 133, 234, 299].

> Filtering is a central core functionality for realizing event-based systems and accurate content delivery. Filtering is performed before delivering a notification to a client or neighboring router to ensure that the notification matches an active subscription from the client or neighbor. Filtering is therefore essential in maintaining accurate event notification delivery.

Filtering increases the efficiency by avoiding the forwarding of notifications to routers that have no active subscriptions for them. Filters and their properties are useful for many different operations, such as matching, optimizing routing, load balancing, and access control. To take some examples, a firewall is an example of a filtering router, and an auditing gateway is a router that records traffic that matches the given set of filters.

Message routing systems can be classified into four categories: channel-based, subject-based, header-based, and content-based. Channel-based systems make the routing decision based on channel names that have been agreed on beforehand by the communicating participants. Subject-based systems make the routing decision based on a single field in the message. Header-based systems use a special header part of the message in order to make the routing decision. Finally, content-based systems use the whole content of the message in making the decision [56].

> In header-based routing, the message has two parts: the header and the body. Only fields in the header are used for making routing and forwarding decisions. Header-based routing is more expressive than subject-based and has performance advantage to content-based routing because only the header of a message is inspected. In content-based routing, the decision is made based on the whole content of a message (the payload). Typically, content-based systems use strongly typed fields in the event message or utilize XML-based document matching.

8.4 Router Configurations

A number of *overlay*-based routing algorithms and router configurations have been proposed. An application layer overlay network is implemented on top of the network layer, and typically overlays provide useful features such as fast deployment time, resilience, and fault-tolerance. An overlay-routing algorithm leverages underlying packet-routing facilities and provides additional services on the higher level, such as searching, storage, and synchronization services.

8.4.1 Basic Configuration

In hierarchical systems, each router has a master and a number of slave routers. Notifications are always sent to the master. Notifications are also sent to slaves that have previously expressed interest in the notifications. The basic hierarchical design is limited in terms of scalability, because one master router is the root of the distribution tree and will receive all the notifications produced in the system.

For acyclic and cyclic topologies, routers employ a different P2P protocol to exchange interest-propagation information and control messages. In this context, the P2P protocol denotes that the topology is not hierarchical. Acyclic topologies allow more scalable configurations than hierarchical topologies, but they lack the redundancy of cyclic topologies. On the other hand, topologies based on cyclic graphs require techniques, such as the computation of minimum spanning trees, to prevent loops and unnecessary messaging.

The hierarchical topology was used in the JEDI system [39, 96], and an acyclic topology with advertisements was used in Rebeca [132, 234, 235]. The Siena project investigated and evaluated the topologies with different interest-propagation mechanisms [53, 57]. In general, the acyclic and cyclic topologies have been found to be superior to hierarchical topologies [39, 53, 237]. The router topology in Gryphon [168, 310] is based on clusters called *cells* and redundant *link bundles* that connect cells. Most researches are focused on

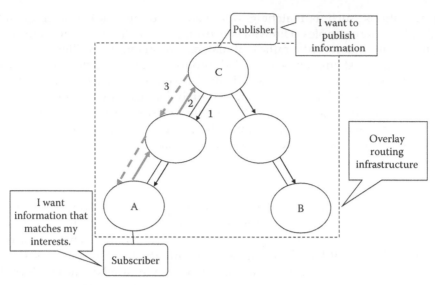

FIGURE 8.5
Example of an overlay routing topology.

static connections between routers; however, a number of dynamic systems have also been proposed [100, 333]—for example, GREEN [297] and REDS [98].

Figure 8.5 illustrates the general content-based routing environment—in which a number of application layer routers offer the interest registration service and maintain routing tables. The depicted topology is acyclic, and the publisher first establishes the publication capability by advertising content. Then subscriptions are connected by using reverse path routing.

8.4.2   Structured DHT-based Overlays

Good overlay routing configuration follows the network-level placement of routers. Many overlays are based on DHTs discussed in Chapter 5, which are typically used to implement distributed lookup structures. Many DHTs work by hashing data to routers/brokers and using a scheme to find the proper data broker for a given data item.

Hermes [256] and Scribe [281] are examples of publish/subscribe systems implemented on top of an overlay network and are based on the rendezvous point routing model. The Hermes routing model is based on advertisement semantics and an overlay topology with rendezvous points. This model was found to compare favorably with the Siena advertisement semantics using an acyclic topology [256].

The rendezvous point model differs from acyclic and cyclic topologies because the routing of a specific type of event is constrained by a special router, the *rendezvous point (RP)*. The RP serves as a meeting point for advertisements and subscriptions and avoids the flooding of advertisements throughout the system.

Rendezvous-based systems limit the propagation of messages using the RP and thus attempt to address scalability limitations presented by the flooding of subscriptions or advertisements. Typically, an RP is responsible for a predetermined event type. RPs may be used to create a type hierarchy. In this case, a message needs to be sent to the proper RP and any supertype RPs, which may increase messaging cost and limit scalability.

8.4.3 Interest Propagation

Notifications are defined using a notification data model, which determines the syntax and structure of events. A filtering model or filtering language determines the syntax, structure, and semantics of filters. These two models in combination represent how information is produced and forwarded to subscribers.

The main functions of a router are to match notifications for local clients and to route notifications to neighboring routers that have previously expressed interest in the notifications. The interest-propagation mechanism is an important part of the distributed system and heart of the routing algorithm. The desirable properties for an interest-propagation mechanism are small routing table sizes and forwarding overhead [237], support for frequent updates, and high performance.

The two well-known operating semantics for content-based pub/sub are the subscription and advertisement semantics. With subscription semantics, the routers propagate subscriptions to other routers, and notifications are sent on the *reverse path* of subscriptions. With advertisement semantics, the routers first propagate advertisements and then, on the reverse path of advertisements, the subscriptions. Notifications are forwarded on the reverse path of subscriptions in both semantics. Advertisements may be used with various routing mechanisms. Advertisements typically have their own routing table, and they are managed using the same algorithms as subscriptions. The removal of an advertisement causes a router to drop all overlapping subscriptions for the neighbor that sent the unadvertisement message. Similarly, an incoming advertisement requires that overlapping subscriptions are forwarded to the neighbor that sent the advertisement message. The use of advertisements considerably improves the scalability of the event system [39, 53, 237].

The four key types of pub/sub routing systems are the following:

- *Simple routing:* Each router knows all active subscriptions in the distributed system, which is realized by flooding subscriptions.

- *Identity-based routing:* A subscription message is not forwarded if an identical message was previously forwarded. This requires an identity test for subscriptions. Identity-based routing removes duplicate entries from routing tables and reduces unnecessary forwarding of subscriptions.

- *Covering-based routing:* A covering test is used instead of an identity test. This results in the propagation of the most general filters that cover more specific filters. On the other hand, unsubscription becomes more complicated because previously covered subscriptions may become uncovered due to an unsubscription.

- *Merging-based routing:* This type of routing allows routers to merge exiting routing entries. Merging-based routing may be implemented in many ways and may be combined with covering-based routing [237]. Also, merging-based routing has more complex unsubscription processing when a part of a previously merged routing entry is removed.

The Siena system was the first system to support both subscription and advertisement semantics and covering-based routing. The Siena system used the notion of covering for three different comparisons: matching a notification against a filter, covering relation between two subscription filters, and overlapping between an advertisement filter and a subscription filter. Covering and overlapping relations have been used in many later event systems, such as Rebeca [235] and Hermes [255, 256]. The combined broadcast and content-based (CBCB) routing scheme extends the Siena routing protocols by combining higher-level routing using covering relations and lower-level broadcast delivery [58]. The protocol prunes the broadcast distribution paths using higher-level information exchanged by routers.

8.5   Siena and Routing Structures

Most research on content-based routing has focused on distributed routing with various semantics or the efficient matching of filters. The routing tables of content-based routers are typically represented as sets. For example, JEDI [96] and Hermes [256] keep filters in a simple table, and Rebeca uses sets and a counting algorithm for finding covering filters and mergeable filters [234]. Two counting-based algorithms are needed for routing, one to determine the covered filters and one to determine the covering filters. A unified approach based on binary decision diagrams (BDDs) has been proposed in [200].

The desirable characteristics for a content-based routing table are efficiency, small size, support for frequent updates, and extensibility and interoperability. The routing table data structure should be generic enough to support a wide range of filtering languages.

> The filters poset data structure was used in the Siena system to store filters by their covering relations and manage information related to forwarded messages. The filters poset can be thought of as the routing table for a Siena router. The poset stores filters by their generality and may also be used to match notifications against filters by traversing only matching filters in the poset, starting from the most general filters. We call the set of most general filters that covers other filters the *root set* of the data structure in question. The root set is also called the *noncovered set* or the *minimal cover set*.

The filters poset is a generic data structure and may be used with various filter semantics, which makes it attractive for dynamic environments. The poset may also be used for various interest-propagation mechanisms, such as subscription and advertisement semantics. On the other hand, this generality has a performance drawback. One of the findings in Siena was that the filters poset algorithm limits the performance of routers and more efficient solutions are needed [58].

A BDD-based routing and matching mechanism was presented in [200]. This approach uses a global predicate index and *modified binary decision diagrams* (MBDs), which are abstract representations of boolean functions. An MBD is used to represent a subscription. They assume typed tuples, and the variables are based on a predefined order. The variable ordering problem is known to be NP-complete. In an MBD-based routing table, publication matching involves iterating the name/value pairs of the event and computing the truth values for the corresponding attribute filters. The attribute filters are located using the predicate index. After this, the MBDs are evaluated using the computed truth values. The cover algorithm involves iterating the MBDs for elements that cover or are covered by the input subscription. If there are no covering elements, the new subscription is a root element. The algorithm stops when a covering element is found. The BDD-based approach can be seen to be more efficient than more generic routing structures such as the poset; however, it assumes global knowledge of predicates and a predefined predicate ordering.

8.5.1   Routing Blocks

A number of generic and modular content-based routing table building blocks can be identified, namely the poset and forest. The routing table can be divided into multiple parts to improve performance and scalability. Figure 8.6 presents three useful routing table configurations that combine the poset and forest structures. The main insight is to separate the routing table into two parts: the external table and the local table—for example, the external table using the poset and the local table using the forest. The term NB in the figure denotes a

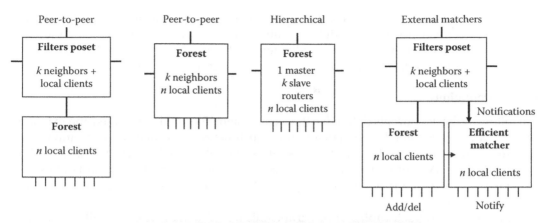

A set of generic building blocks for filter cover–based routing.
Can be extended with optimizations such as pruning and caching.

FIGURE 8.6
Routing blocks for covering-based routing.

neighbor interface. The forest is used to maintain client subscriptions (and advertisements), and the poset is updated only when the root set of the forest changes. The Siena system also uses the poset to store local filters, which is not efficient. Assuming that there exist covering relations between local filters, this separation ensures that the external table is not burdened with frequent updates by local clients.

In Siena, the subscriptions of a local client are not handled independently of each other. Any filter from a local client that is covered by a new filter from the same client will be removed. Similarly, when a client removes a filter (unsubscribes), any filters that are covered by the removed filter will be removed. This approach requires that clients are able to compute the covering relations between their filters and explicitly manage their filter sets. In addition, in this model it is not possible to transparently change the routing algorithm semantics without making changes to the client code.

A separate data structure called the poset-derived forest can be used to manage filters from local clients. In this case, client-side filter set management is simple and efficient. The second benefit is increased performance, because the forest supports faster insertions and deletions than the poset. This model also supports extending the system to support filter merging (also called aggregation and summarization)—for example, by starting with the root set of the forest storing the local clients.

The figure also illustrates use of the forest as both local and external routing tables. This configuration is feasible when there are many local clients, but the external forwarding is more complicated than for the poset. The forest can also be used for hierarchical routing with the master and slave interfaces identified. The forest may also store local clients. The separation into two parts allows for prioritizing operations.

A more efficient matching data structure may be introduced into the filter-based routing core. In this case, any addition (*add*) and deletion (*del*) operations by local clients are processed by the forest and also reflected to the efficient matcher. Only the root set is updated to the poset, which is the external routing structure. When an incoming event matches the local interface (root filters of the forest), the notification is sent to the efficient matcher.

With hierarchical and P2P routing, the routing blocks may be used to enhance rendezvous-based routing models, such as Hermes. In the Hermes model with filters, advertisements are always propagated toward the RP. Subscriptions are propagated toward the RP and towards any overlapping advertisements. Therefore, advertisements may be stored using a forest and subscriptions using a forest or a poset. In both cases local clients are stored using a forest. For rendezvous-based models, the subscription poset must be extended to support any subscriptions that should be forwarded toward the RP. This is accomplished by using a virtual advertisement from the RP that covers all subscriptions of the designated type.

8.5.2 Definitions

We follow the basic concepts defined in the Siena system [55] and later refined and extended in Rebeca [234]. A filter F is a stateless Boolean function that takes a notification as an argument. Many event systems use the operators of Boolean logic, *AND*, *OR*, and *NOT*, to construct filters. A filtering language specifies how filters are constructed and defines the various predicates that may be used. A predicate is a language-specific constraint on the input notification.

A filter is said to match a notification n if and only if $F(n) = true$. The set of all notifications matched by a filter F is denoted by $N(F)$. A filter F_1 is said to cover a filter F_2, denoted by $F_1 \sqsupseteq F_2$, if and only if all notifications that are matched by F_2 are also matched by F_1—i.e., $N(F_1) \supseteq N(F_2)$. We also say that F_1 has *equal or greater selectivity* than F_2. Similarly, F_2 has *equal or lesser selectivity* than F_1. The filter F_1 is equivalent to F_2, written $F_1 \equiv F_2$, if $F_1 \sqsupseteq F_2$ and $F_2 \sqsupseteq F_1$. The filter F_1 is *incomparable* with F_2, if $F_1 \not\sqsupseteq F_2$ and $F_2 \not\sqsupseteq F_1$. The $\sqsupseteq$ relation is reflexive and transitive and defines a partial order.

A set of n filters $S_F = \{F_1, \ldots, F_n\}$ covers a filter F_k if and only if $N(S_F) \supseteq N(F_k) \Leftrightarrow \bigcup_i^n N(F_i) \supseteq N(F_k)$. Covering of two sets follows from this.

An advertisement A is said to overlap with the subscription S, denoted by $A \simeq S$, when their filters overlap. Two filters, F_1 and F_2, are overlapping if and only if $N(F_1) \cap N(F_2) \neq \emptyset$.

As an example, we can consider three filters using the notation *(filter, constraint)*: $(F_1, x < 10)$, $(F_2, x \in [5, 9])$, and $(F_3, x \in [8, 15])$. The constraints are defined for the variable x over integers. We have $F_1 \sqsupseteq F_2$, since the range $[5, 9]$ is contained in $x < 10$. We have $F_1 \not\sqsupseteq F_3$, because the range $[8, 15]$ is not totally contained in $x < 10$. It is also clear that the ranges do not contain each other, hence $F_2 \not\sqsupseteq F_3$ and $F_3 \not\sqsupseteq F_2$. On the other hand, it is clear that $F_1 \simeq F_2$. Also $F_1 \simeq F_3$ since $x < 10$ and $[8, 15]$ overlap.

8.5.3 Siena Filters Poset

The filters poset data structure was used in the Siena-distributed event system for maintaining covering relations between filters [55]. In Siena's P2P configurations, the poset stores additional information for each subscription that is inserted into the poset. The *subscribers(f)* set gives the set of subscribers for the given subscription filter f, and, similarly, *forwards(f)* contains the subset of peers to which f needs to be sent. Algorithm 8.1 presents the steps needed to process a subscription *subscribe(X, f)* where X is the subscriber and f is the filter representing the subscription [55, 57].

In distributed operation based on an acyclic graph router topology, the Siena server defines the set *forwards(f)* as presented in the equation

$$forwards(f) = neighbors - NST(f) - \bigcup_{f' \in P_s \wedge f' \sqsupseteq f} forwards(f'). \tag{8.2}$$

The *neighbors* set contains the event brokers connected to the current broker (one application-level hop distance). The functor *NST* (not on any spanning tree) means that the propagation of f must follow the computed spanning trees rooted at the original subscribers of f. With acyclic topologies, *NST* contains the neighbor that sent f. P_s denotes the subscription poset. Using the equation, f is never forwarded to the neighbor that sent it. Due to the last term of the equation, the subscription is not forwarded to any routers that have already been sent a covering subscription.

Algorithm 8.1 Filter processing in the subscription *subscribe(X, f)*

Function: *subscribe(X, f)*

1. If a filter f' is found for which $f' \sqsupseteq f$ and $X \in$ *subscribers(f')*, then the procedure terminates, because f for X has already been subscribed by a covering filter.

2. If a filter f' is found for which $f' \equiv f$ and $X \notin$ *subscribers(f')*, then X is added to *subscribers(f')*. The server removes X from all subscriptions covered by f. Also, subscriptions with no subscribers are removed.

3. Otherwise, the filter f is placed in the poset between two possibly empty sets: immediate predecessors and immediate successors of f. The filter f is inserted and X is added to *subscribers(f)*. The server removes X from all subscriptions covered by f, and subscriptions with no subscribers are also removed.

Because X is removed from all subscriptions covered by f, an intermediary server does not know which subscriptions should be forwarded due to unsubscription. This information is essentially lost by this optimization; however, the origin of the subscriptions has this information and propagates any subscriptions due to the unsubscription in the same message, which is applied atomically by other servers. The *unsubscribe(X, f)* removes X from the *subscribers* set of all subscriptions that are covered by f. Filters with empty subscriber sets are removed. Algorithm 8.2 gives an outline of subscription processing. The model may be extended with advertisements [57].

The message-forwarding behavior of hierarchical routing is simple. This behavior becomes more complex when a router has multiple neighboring routers. Siena uses the *forwards* set to compute destinations for messages in P2P routing.

The *forwards(f)* set is determined using Equation (8.2). The last term of the equation means that the removal of an entry in a *forwards* set may affect the *forwards* sets of other subscriptions. This happens during unsubscriptions and may require some of the uncovered

---

Algorithm 8.2 Message handlers for subscription semantics.

Function: IncomingSub(f,*source*)

1. Add (f,*source*) to P_s.

2. Forward subscription message using *forwards(f)* to any new neighbors in the set.

Function: IncomingUnsub(f,*source*)

1. Remove (f,*source*) from P_s.

2. Let F_O denote the old forwards set and F_N a newly computed forwards set for f after the subscriber *source* has been removed from the *subscribers* set. If the *subscribers* set is empty, then $F_N = \emptyset$. The unsubscription is forwarded to $F_O \setminus F_N$. The set may be empty if there are subscriptions from other neighbors that cover f. The *forwards* sets of subscriptions covered by f may change, which may require the forwarding of new subscriptions. Any uncovered subscriptions in P_s are forwarded with the unsubscription message. An uncovered subscription is such that its forwards set gains an additional element due to the removal of a covering filter.

---

subscriptions to be forwarded. Only elements in the root set or the direct successors of elements in the root set may have a nonempty *forwards* set [322, 324].

8.5.4 Advertisements

The basic subscription semantics may be optimized by using advertisements. In this model, advertisements are propagated to every node, and subscriptions are propagated only toward advertisers that have previously advertised an overlapping filter. The idea is to use the additional advertisement information to prevent subscription flooding. The model uses two poset data structures, one for each type of message. Since the poset-derived forest can be made equivalent to the filters poset, it is also a useful data structure for advertisement semantics. Advertisements from local clients can be stored in a redundant forest.

In advertisement semantics, a second poset P_a is used for advertisements [55]. The sets *advertisers(a)* and *forwards(a)* are needed for each advertisement $a \in T_A$, where T_A is the set of all advertisements in the poset. Instead of forwarding subscriptions to a global set *neighbors*, a set constrained by advertisements is used as presented by the equation

$$neighbors_s = \bigcup_{a \in T_A : a \simeq s} advertisers(a) \cap neighbors. \tag{8.3}$$

In this case, Equation (8.2) uses the $neighbors_s$ set instead of the *neighbors* set. An advertisement may thus result in a number of subscriptions being forwarded to the sender of the advertisement. The process of unadvertisement is similar to unsubscription. Algorithms 8.3 and 8.4 give an outline of message processing with advertisement semantics. The algorithms are derived from [55] and [234].

8.5.5 Poset-derived Forest

The poset-derived forest data structure is used to store filters by their covering property with other filters [322, 324], and it offers linear time processing for both insertions and deletions instead of superlinear time of the filters poset. Moreover, the space requirement is linear, which contrasts the superlinear space required by the poset to store both the nodes and the edges between them.

Algorithm 8.3 Subscription message handlers for advertisement semantics.

Function: IncomingSub(f,*source*)

1. Add (f,*source*) to P_s.
2. Calculate $neighbors_s$ using P_a and Equation (8.3).
3. Send subscription message to *forwards(f)*.

Function: IncomingUnsub(f,*source*)

1. Remove (f,*source*) from P_s.
2. Forward unsubscription following the procedure in Algorithm 8.2. The set may be empty if there are subscriptions from other neighbors that cover f. The *forwards* sets of subscriptions covered by f may change, which may require the forwarding of new subscriptions. An uncovered subscription is such that its forwards set gains an additional element due to the removal of a covering filter.

Algorithm 8.4 Subscription message handlers for advertisement semantics.

Function: IncomingAdv(a, *source*)

1. Add (a, *source*) to P_a.

2. Forward advertisement message to *forwards(a)*.

3. Determine the set of overlapping subscriptions using P_s for which a is the only advertisement from the *source* that overlaps and send them to the *source*. In other words, any subscriptions that have not yet been sent are forwarded to the advertising node (*source*). Those subscriptions that overlap with an existing advertisement from the *source* have already been forwarded, so they are not processed. The overlapping set is found by iterating over the first two levels of P_s and testing the overlap of subscriptions with the advertisement.

Function: IncomingUnadv(a, *source*)

1. Remove (a, *source*) from P_a.

2. Forward unadvertisement in a similar fashion that the unsubscription is forwarded. The *forwards(a)* set may be empty if there are advertisements that cover a from other neighbors. Forward any uncovered advertisements in P_a.

3. Remove any subscriptions for *source* that are no longer needed. All subscriptions are removed from neighbors other than the *source* that do not have an associated overlapping advertisement from some other neighbor.

A pair $(\mathcal{F}, \succ)$ represents the poset-derived forest, where $\mathcal{F}$ is a finite set of filters and $\succ$ is a subset of the covering relation. More formally:

DEFINITION 8.1
A pair $(\mathcal{F}, \succ)$ is a poset-derived forest with base set $\mathcal{F}$, if

1. *$\mathcal{F}$ is a finite set of filters and $\succ$ is a relation between filters in $\mathcal{F}$.*

2. *For each $a \in \mathcal{F}$ there is at most one $b \in \mathcal{F}$ for which $b \succ a$, i.e., $(\mathcal{F}, \succ)$ is a forest with the relation $\succ$ going from parent to child.*

3. *If $a, b \in \mathcal{F}$ and $b \succ a$, then $b \sqsupseteq a$.*

It is convenient for uniformity of treatment to imagine the roots of the trees belonging to $(\mathcal{F}, \succ)$ to be children of a node not in $\mathcal{F}$, which we will call the *imaginary root* of $(\mathcal{F}, \succ)$.

$(\mathcal{F}, \succ)$ is called *maximal in $\mathcal{F}$* if there does not exist $a, b \in \mathcal{F}$ for which $(\mathcal{F}, \succ \cup\{(a, b)\})$ is a poset-derived forest. It is clear that any poset-derived forest can be extended to a maximal one by adding pairs to the relation $\succ$.

In applications we typically require the maximality criterion to hold. The maximality criterion may be generalized to apply at any level of the forest, which is called *sibling-purity*. Sibling-purity at a node means that the node's children in the forest are incomparable with each other. In other words, a sibling-pure forest ensures that nodes are locally placed as far away from the root nodes as possible.

The *add* and *del* operations are simple and efficient for the forest. *Add* is based on a depth-first search on F to find a suitable parent for the new node. *Del* simply removes a node and performs *add* for children as subtrees starting from the current parent. The algorithms

---

Algorithm 8.5 *Add* and *del* procedures for the forest.

Let $(\mathcal{F}, \succ)$ be a poset-derived forest. It is assumed that there is an efficient way to find a node in $\mathcal{F}$ based on its identifier. In subsequent examination, references to "larger" and "smaller" are to be taken with respect to the relation $\sqsupseteq$. We define the following algorithms with inputs $\mathcal{F}$ and a filter x and output a poset-derived forest:

add($\mathcal{F}$, x): This algorithm maintains a *current node* during its execution. First, set the current node to be the imaginary root of $\mathcal{F}$.

1. If x is already in the forest, return without changes.

2. Else if x is incomparable with all children of the current node, add x as a new child of the current node.

3. Else if x is larger than some child of the current node, move all children of the current node that are smaller than x to be children of x and make x a new child of the current node.

4. Else pick a child of the current node that is larger than x, set the current node to this picked child and repeat this procedure from step 2.

del($\mathcal{F}$, x): Let C be the set of children of x and r be the parent of x. Then run add for each of the elements of C starting from step 2 and setting r as the current node. In this an element of C carries the whole subtree rooted at it with the addition. To preserve sibling-purity, any siblings of a relocated node that are smaller than the node must be relocated deeper into the tree using add.

---

are presented in more detail in [322, 324]. The filters poset and the poset-derived forest compute the minimal cover set for the input set (Definition 8.2).

DEFINITION 8.2
A minimal cover set or a root set of the filters poset or poset-derived forest is a set R such that there does not exist an element $a \in R$ for which $b \sqsupseteq a$ and $b \in R$.

Sibling-purity is very easy to maintain for the add operation but more complicated for the del operation. It is expected that, for some application areas, such as hierarchical routing or the management of filters from local clients, it is not necessary to maintain sibling-purity for the del operation. This simplifies the del operation in Algorithm 8.5.

Algorithm 8.5 assumes that there is an efficient way to find if a filter has already been placed into the structure. This is possible using *syntactic equivalence* using hashtables. In syntactic equivalence, canonical representations of filters are compared. Syntactic equivalence is not necessarily implied by *semantic equivalence*—e.g., $F_1 \sqsupseteq F_2 \wedge F_2 \sqsupseteq F_1$. Semantic equivalence is computationally more complex to determine, whereas syntactic equivalence may be achieved in constant or near-constant time, and it detects all semantically identical filters with simple filtering languages. We note that this restriction to syntactic equivalence does not break the data structure or the routing algorithms. Filters that fail the equivalence testing will simply be placed into the structure.

8.5.6 Filter Merging

Filter merging is a technique to find the minimum number of filters that represent a set of subscriptions defined in the content space. Filter merging approaches this by fusing and combining the filters using logical rules. Filter covering is a related technique, which is used to remove filters that are covered by other, more general, filters. Merging and covering are needed to reduce processing power and memory requirements both on client devices and on event routers. These techniques are typically general and may be applied to subscriptions, advertisements, and other information represented using filters.

A *false negative* is an event that was not matched and delivered when it should have been. A *false positive* is a message that was matched, but it should not have been. In publish/subscribe, false negatives should never occur, and they indicate a serious error in the system. False positives may occur, and it is possible to balance between efficiency and accuracy.

A filter-merging-based routing mechanism was presented in the Rebeca-distributed event system [234]. The mechanism merges conjunctive filters using perfect merging rules that are predicate-specific. Merging was used only for simple predicates in the context of a stock application [234, 237].

The optimal merging of filters and queries with constraints has been shown to be NP-complete [94] in the multicast environment. This work considered query merging for allocating query answers to multicast channels. Subscription partitioning and routing in content-based systems have been investigated in [344, 345] using Bloom filters [32] and R-trees [33] for efficiently summarizing subscriptions.

Bloom filters investigated in Chapter 7 are an efficient mechanism for probabilistic representation of sets and support membership queries, but they lack the precision of more complex methods of representing subscriptions. Bloom filters and additional predicate indices were used in a mechanism to summarize subscriptions [327, 328]. An arithmetic attribute constraint summary (AACS) and a string attribute constraint summary (SACS) structure were used to summarize constraints because Bloom filters cannot directly capture the meaning of other operators than equality. The subscription summarization is similar to filter merging, but it is not transparent because routers need to be aware of the summarization mechanism. Filter merging, on the other hand, does not necessarily require changes to other routers. The event routers need to be aware of the summarization mechanism. In addition, the set of attributes needs to be known a priori by all brokers, and new operators require new summarization indices. The benefit of the summarization mechanism is improved efficiency, since a custom-matching algorithm is used that is based on Bloom filters and the additional indices.

A BDD-based matching algorithm was proposed in [46]. A BDD-based merging algorithm was presented in [200]. The exact rules for dynamic filter merging were not elaborated in this work. The algorithm removes all subscriptions, which are covered by a new merger. This requires that all routers are aware of the merging technique in order to support safe unsubscriptions.

A general model for filter merging for Siena-style content-based routers was presented in [323]. Filter merging may be applied in different places in the event router. The three key merging scenarios and techniques are *local merging*, *root-merging*, and *aggregate merging*. In the first scenario, filter merging is performed within a data structure. In the second scenario, filter merging is performed on the root sets of local filters, edge/border routers, and hierarchical routers. In the third scenario, filter merging is performed on the two first levels of a peer-to-peer data structure, such as the filters poset. The latter two scenarios are examples of remote merging [323].

Figure 8.7 presents two router configurations with filter merging and highlights the modular structure of content-based routers. Local clients are stored by the forest data structure, which is a good candidate structure for storing filters from local clients. The key idea is to perform merging separately for the external routing table and structure that stores filters from local clients. Moreover, given that a cover-based structure is used, it is sufficient to test merging candidates only from the two first levels of the structure. In many cases, it is sufficient to scan the root set or parts of it to find merging opportunities.

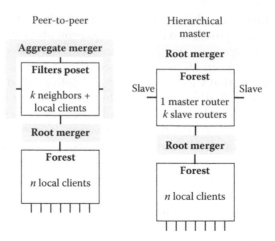

FIGURE 8.7
Routing blocks for merging-based routing.

8.6   Hermes

Hermes [255, 256] is a peer-to-peer event system based on an overlay called Pan that supports a variant of the advertisement semantics. Hermes leverages the features of the underlying overlay system for message routing, scalability, and improved fault-tolerance. Hermes supports the basic pub/sub operations introduced previously. Rendezvous points are used to coordinate advertisement and subscription propagation. The RP manages an event type and Hermes supports chaining RPs into type hierarchies. The RP of an event type is obtained by hashing the event type to the flat addressing space of the overlay [256].

A *rendezvous point* is a special node in the routing network that is used to coordinate signaling. Rendezvous points are used in many overlay routing systems [280, 308, 363] to reduce communication costs and realize nonfixed indirection points. The rendezvous points are uniformly distributed over the addressing space. The placement of event types (RPs) using uniform distribution is motivated by the fact that the types are disjoint from the viewpoint of matching. On the other hand, event traffic distribution may well be nonuniform, which should also be taken into account. The problem with nonuniform traffic distribution is that an RP may be located on the other side of the network. The RP may then become a performance and scalability bottleneck.

Hermes rendezvous points are established using a special message that establishes an event type to a rendezvous point, which owns the address of the hashed type identifier. Event type conforms to a schema that the client software may request using an API call. This is required for type-safe subscriptions.

Hermes supports two routing algorithms: *type-based routing* and *type/attribute-based routing*. In type-based routing, all messages are propagated toward the RP: subscriptions, advertisements, and notifications. Type-based routing does not support filtering, but compares events based on their type. Subscriptions and advertisements are local to a branch of the multicast tree rooted at the RP, and they are not forwarded by the RP. This means that notifications are always to be sent to the RP. Type/attribute-based routing is similar to type-based routing but supports filtering with covering relations and, instead of sending all notifications to the RP, notifications are sent on the reverse path of subscriptions. In this case, advertisements are sent only to the RP. Subscriptions are always sent on the reverse path of advertisements. The RP forwards subscription messages to overlapping advertisements. The type/attribute-based routing is more suitable to scenarios where event traffic is not uniformly distributed, because notifications are not always sent to the RP.

One key feature of Hermes is connecting RPs into type hierarchies. In *subscription inheritance routing*, advertisements are sent only to the RP that maintains the event type. Subscriptions are forwarded by the RP to all RPs with descendant types. In *advertisement inheritance routing*, the RP forwards the advertisement recursively to all RPs of all ancestor event types. Also, notifications are forwarded to all ancestor event types, because they are sent on the same forward path as advertisements.

Hermes uses heartbeat messages to detect server and RP failures. The underlying overlay allows location of a new server that takes over the responsibilities of a failed node. Routing tables are simply sent toward the RP, and the overlay will provide a new route with a new server. Hermes supports RP replication by synchronizing advertisement and subscription status between different replicas. The replicas are placed in the same multicast tree to avoid overhead due to message propagation. Load balancing of traffic between RPs is not discussed.

Figure 8.8 illustrates rendezvous-point-based operation using 11 steps: 1. A publisher advertises an event type (and a filter in type/attribute-based routing). 2. The advertisement is forwarded to the rendezvous point. 3. A subscriber subscribes to an event of the same type (and a filter in type/attribute-based routing). 4. The subscription message is not covered (type or filter) at any intermediate broker and is forwarded to the rendezvous point. 5. Another subscriber subscribes. 6. The subscription message is propagated toward the RP. 7. The publisher publishes an event. 8–11. The message is sent using the multicast tree rooted at the RP.

In type-based routing, any events conforming to the advertisement from the publisher are sent on the forward path of the advertisement to the RP, which then forwards the event on the reverse path of any subscriptions. In type/attribute-based routing, the RP sends the subscriptions on the reverse path of advertisements. Any events conforming to the advertisement from the publisher are sent on the reverse path of subscriptions.

The model used by the Hermes system is the familiar advertisement semantics model, with three key differences:

- All messages (type-based routing) or advertisements and subscriptions (type/attribute-based routing) are sent toward the RP. Thus routing topology is constrained by the RP.

- Advertisements are introduced only on the path from the advertiser to the RP.

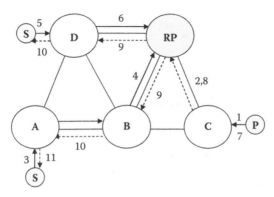

FIGURE 8.8
Rendezvous-point-based routing in Hermes.

- Subscriptions are introduced on the path from the subscriber to the RP. In addition, for type/attribute-based routing, subscriptions are sent on the reverse path of any overlapping advertisements.

These differences are interesting because advertisement becomes a local property of a branch of the multicast tree rooted at an RP. This may be modeled using virtual advertisements. In this case, an RP has virtual advertisements for all events of the event type managed by the RP and hence subscriptions are sent toward it. In the following examination we assume that the overlay topology is static; a dynamic topology would require a more complex investigation.

8.7   Formal Specification of Content-based Routing Systems

8.7.1   Valid Routing Configuration

The valid routing configuration determines that the publish/subscribe system does not manifest illegal traces. A trace is a sequence of operations, such as subscribe, notify, and unsubscribe. Any valid routing configuration must satisfy the following constraints on traces presented using the operators of the *linear temporal logic (LTL)*. LTL formulas are used to define a specification, and a system is correct when it exhibits only traces allowed by the specification. $\square$ denotes "always", $\lozenge$ "eventually", and $\bigcirc$ "next".

Property 8.1 gives the liveness constraint for the basic publish/subscribe system with subscription semantics. The liveness property defines when a notification should be delivered and ensures that notifications are eventually delivered. Property 8.2 gives the safety constraint, which ensures that incorrect events are not processed and delivered. The properties are from the definitions in [234], with minor changes in presentation.

Property 8.1
Liveness:

$$\square[\ Sub(A,F) \Rightarrow [\lozenge\square(\ Pub(B,n) \wedge n \in N(F) \Rightarrow \lozenge\ Notify(A,n))] \vee$$

$$[\lozenge Unsub(A,F)]],$$

specifies that a subscription with filter F and the publication of an event n that matches the subscription will lead to an eventual notification of subsequent publications of that event unless the subscription is invalidated by unsubscription.

Property 8.2
Safety:

$$\square[Notify(A, n) \Rightarrow [\bigcirc\square\neg Notify(A, n)] \wedge$$

$$[n \in Published] \wedge$$

$$[\exists F \in Subs(A) : n \in N(F)]],$$

specifies that a notification is delivered only once, that it has been published previously, and that the recipient has a matching subscription. Published is the set of published events, and the set Subs gives the subscriptions for each client.

Since it may be difficult to maintain these properties in dynamic pub/sub systems, they may be relaxed. A self-stabilizing pub/sub system ensures correctness of the routing algorithm against the specification and convergence [234]. The safety property may be modified to take self-stabilization into account by requiring eventual safety. The safety and liveness properties were extended in [321] with the notion of *message-completeness* and using propositional temporal logic. A message-complete pub/sub system eventually acknowledges subscriptions and guarantees the delivery of notifications matching acknowledged subscriptions.

8.7.2 Weakly Valid Routing Configuration

The weakly valid routing configuration guarantees only the delivery of notifications to those subscriptions whose update process has terminated. A routing algorithm that uses the weakly valid routing configuration and ensures that every update process terminates satisfies Properties 8.1 and 8.2 [234].

We call all update procedures that have ended successfully complete in the topology and use completeness to characterize and prove properties of pub/sub mobility. By topology, we mean the logical network among brokers that is used to route messages. Typically, the topology for advertisements consists of the logical connections between the brokers, and for subscriptions it is constrained by advertisements.

The completeness of subscriptions and advertisements is given by Definition 8.3. Advertisements are complete when they have been propagated to every node that may issue an overlapping subscription in the future. Similarly, subscriptions are complete when they have been introduced at every node that has an overlapping advertisement. This formulation is flexible enough to be useful for various routing protocols. Completeness may be used to characterize the whole routing system. In addition, it may also be used to characterize a part of the routing system, such as a *path*.

DEFINITION 8.3
An advertisement A is complete in a pub/sub system PS if there does not exist a broker r with an overlapping subscription that has not processed A. Similarly, a subscription S is complete in PS if there does not exist a broker r such that r has an advertisement that overlaps with S and S is not active on r.

8.7.3 Mobility-Safety

In distributed pub/sub systems it is evident that, after issuing a subscription, it will take some hops before the subscription is activated for all publishers. During this time several notifications may be missed. In the mobility-aware weakly valid routing configuration, false negatives that occur during topology reconfiguration caused by subscriptions and advertisements from stationary components are tolerated. False negatives that occur during client mobility are not tolerated. A mobility-safe pub/sub system can be defined as follows:

DEFINITION 8.4
A pub/sub system is mobility-safe if, starting from an initial configuration C_0 at time T_0 and ending in a configuration C_e at time T_e, handovers (mobile clients) will not cause any false negatives.

8.8 Pub/sub Mobility

Event systems have traditionally focused on event dissemination in the fixed network, where clients are stationary and have reliable, low-latency, and high bandwidth communication links. Recently, mobility support and wireless communication have become active research topics in many research projects [97, 165, 166, 258, 259] working with event systems, such as Siena [57] and Rebeca [131, 236].

Mobility support [165, 166, 259] is a relatively new research topic in event-based computing. Mobility is an important requirement for many application domains, where entities change their physical or logical location. Mobile IP is a layer-3 mobility protocol for supporting clients that roam between IP networks [175, 252]. Higher-level mobility protocols are also needed in order to provide efficient middleware solutions—for example, session initiation protocol (SIP) mobility [287] and wireless CORBA [245]. Event-based systems require their own mobility protocols in order to update the event-routing topology and optimize event flow.

In order to understand event-routing, we need to have useful metrics to characterize the system. Besides message complexity and computing power, the two most important metrics are the number of false positives and negatives. *False positives* are events that are delivered but were not subscribed, and, similarly, *false negatives* are events that were subscribed but were not delivered upon publication. Clearly, the presence of false negatives indicates a serious error in any event system. Therefore, we are interested in proving that a candidate event system does not manifest this erroneous behavior.

Intuitively, given that we first establish a new flow and only after the successful completion of this tear down the old one, there should not be any false negatives, which would satisfy the requirement for mobility-safety. A perfect topology update protocol may be described using flooding that delivers all events to all brokers. This naive protocol also ensures that mobile components will receive all events that match their filters, albeit with a high cost in false positives. A good mobility protocol is mobility-safe, minimizes the number of false positives, and minimizes the signaling cost.

Recently, mobility extensions have been presented for several well-known distributed event systems, such as Siena and Rebeca. JEDI was one of the early systems to incorporate support for mobile clients with the move-in and move-out commands [97]. JEDI maintains causal ordering of events and is based on a tree topology, which has a potential performance bottleneck at the root of the tree with subscription semantics. Elvin is an event system that supports disconnected operation using a centralized proxy but does not support mobility between proxies [316].

Siena is a scalable architecture based on event routing that has been extended to support mobility [49–51]. The extension provides support for terminal mobility on top of a routed event infrastructure. In addition, the Rebeca event system supports mobility in an acyclic event topology with advertisement semantics [357].

Rebeca supports both logical and physical mobility. The basic system is an acyclic routed event network using advertisement semantics. The mobility protocol uses an intermediate node, between the source and target of mobility, called junction for synchronizing the servers. If the brokers keep track of every subscription, the junction is the first node with a subscription that matches the relocated subscription propagated from the target broker. If covering relations or merging is used, this information is lost, and the junction needs to use content-based flooding to locate the source broker [235].

JECho is a mobility-aware event system that uses opportunistic event channels in order to support mobile clients [73]. The central problem is to support a dynamic event delivery topology, which adapts to mobile clients and different mobility patterns. The requirements are addressed primarily using two mechanisms: proactively locating more suitable brokers and using a mobility protocol between brokers; and using a load-balancing system based on a central load-balancing component that monitors brokers in a domain. The topology update and its mobility-safety are not discussed.

Mobility support in a generic routed event infrastructure, such as Siena and Rebeca, is challenging because of the high cost of the flooding and issues with mobile publishers. The standard state transfer protocol consists of four phases:

1. Subscriptions are moved from broker A to broker B.
2. B subscribes to the events.
3. A sends buffered notifications to B.
4. A unsubscribes if necessary.

The problem with this protocol is that B may not know when the subscriptions have taken effect—especially if the routing topology is large and arbitrary. This is solved by synchronizing A and B using events, which potentially involves flooding the content-based network.

Recent findings on the cost of mobility in hierarchical routed event infrastructures that use unicast include that network capacity must be doubled to manage with the extra load of 10% of mobile clients [42]. Recent findings also present optimizations for client mobility: *prefetching, logging, home-broker*, and *subscriptions-on-device*. Prefetching takes future mobility patterns into account by transferring the state while the user is mobile. With logging, the brokers maintain a log of recent events and only those events not found in the log need to be transferred from the old location. The home-broker approach involves a designated home broker that buffers events on behalf of the client. This approach has extra messaging costs when retrieving buffered events. Subscriptions-on-device stores the subscription status on the client so it is not necessary to contact the old broker. In this study the cost of reconfiguration was dominated by the cost of forwarding stored events (through the event routing network).

The cost of publisher mobility has also been recently addressed [240, 241]. They start with a basic model for publisher mobility that simply tears down the old advertisement and establishes it at the new location after mobility. Thus a specific handover protocol is not needed. They confirm the high cost of publisher mobility and present three optimization techniques, namely *prefetching, proxy*, and *delayed*. The first exploits information about future mobility patterns. The second uses special proxy nodes that advertise on behalf of the publisher and maintain the multicast trees. The third delays the unadvertisement at the source to exploit the overlap of advertisements but does not synchronize the source and target brokers. The publisher mobility support mechanisms used in the study are not necessarily mobility-safe.

The Siena event system was extended with generic mobility support, which uses existing pub/sub primitives: publish and subscribe [50, 51]. The mobility-safety of the protocol was formally verified. The benefits of a generic protocol are that it may work on top of various pub/sub systems and requires no changes to the system API. On the other hand, the performance of the mobility support decreases, because mobility-specific optimizations are difficult to realize when the underlying topology is hidden by the API. Indeed, in this section we show that a general API-based pub/sub mobility support may have a very high cost in terms of message exchanges.

The Siena generic mobility support service, the *ping/pong protocol*, is implemented by proxy objects that reside on access routers. Figure 8.9 presents an overview of the process: (1) the client arrives to access point B from A and sends the *move-in* request to the new local proxy, (2) a ping request is sent, and (3) a response will be received eventually from

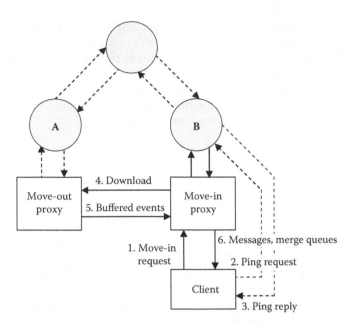

FIGURE 8.9
Generic mobility in pub/sub.

the old proxy. The response can also be called a *pong*. The pong message ensures that subscriptions are fully propagated from *B* to *A*. (4) the client sends a *download* request for buffered events and (5) the buffered events are sent to the proxy. Finally, in (6) the client receives the messages and duplicates are removed.

The following guidelines have been proposed for engineering mobility-safe pub/sub systems [325]:

- The generic protocol is mobility-safe and applicable to various underlying pub/sub systems, but it is very inefficient and does not allow pub/sub system or topology-specific optimizations. The mobility-safety of this mechanism requires that the ping/pong interaction is sufficient to establish the completeness of the path or paths.

- The general acyclic graph-routing topology is more efficient than the generic protocol but suffers from the problem that the source broker needs to be located using event routing. Since covering and merging do not preserve information pertaining to the original broker that issued a subscription or advertisement, the use of content-based flooding may be required. This routing topology also suffers from the incompleteness of subscriptions, and thus the covering optimization that uses out-of-band communication cannot be performed if mobility-safety is required. Incompleteness may also cause a broker to flood subscriptions to several exit interfaces. Incompleteness of the subpath from the source broker to the destination broker may be corrected, but it has a high cost due to potential content-based flooding.

- Rendezvous point models with cyclic overlay routing support better coordination of mobility. With rendezvous points, advertisements are no longer flooded throughout the network, which improves update latency and performance. Moreover, rendezvous points may be used for fast completeness checks. The covering optimization may be used with completeness checking. Furthermore, the overlay address may be used to prevent content-based flooding by consulting the over-lay routing tables and finding the proper next hop. On the other hand, the upper bound cost for cyclic topologies may be higher than for general acyclic graphs if the moving subscriber has subscribed to multiple rendezvous points that have to be updated.

* Rendezvous point models with acyclic overlay routing have the simplifying fea-tures mentioned above and the upper bound cost cannot be greater than for the general acyclic graphs.

Mobility can be improved by using the following techniques: overlay-based routing, rendezvous points, and completeness checking. Overlay addresses prevent the content-based flooding problem. This abstracts the communication used by the pub/sub system from the underlying network-level routing and allows the system to cope with network-level routing errors and node failures. Rendezvous points simplify mobility by allowing better coordination of topology updates. There is only one direction where to propagate updates for a single rendezvous point. Completeness checking ensures that subscriptions and advertisements are fully established (complete) in the topology. This is needed to perform the covering optimization.

9

Security

Given the scalable and flexible distribution solutions enabled by peer-to-peer (P2P) and overlay technologies, we are faced with the question of security risks. The authenticity of data and content needs to be ensured, also taking required levels of anonymity, availability, and access control into account. This chapter examines the security challenges of P2P and overlay technologies and then outlines a number of solutions to mitigate the examined risks. Issues pertaining to identity, trust, reputation, and incentives need to be analyzed as well.

The most visible security challenge pertains to the authentication of data and content distributed in the networks. This is also the risk sector that has the largest number of concerned parties facing possible losses and other kinds of adverse effects resulting from inadequate digital defenses. The need for strong anonymity is closely linked to the content integrity, and they can often be considered as technically unseparable challenges for design efforts. Lesser interlinked problems are related to availability and sufficient access control.

9.1 Overview

The security challenges most often appear as malicious attacks: realized or threatened. There are many types of these attacks, and the mutability of attack methods is very high, producing an ever growing set of new risks when the underlying networks are changed in any way. Strengthening the networks against new challenges is also an alteration bound to bring forth yet new methods of malicious attacks; thus the cycle of security solution/malicious attack is in practice never-ending.

The malicious schemes are multifarious, and they often use ingenious methods for attacking networks. Possible attack scenarios are presented below:

- Attacker controls malicious nodes masquerading as legal nodes.
- Attacker floods DHT with data.
- Attacker returns incorrect data.
- Attacker denies data exists or supplies incorrect routing info.
- Attackers may seek a quorum in k-redundant networks.

The methods normally used within overlay networks to combat threats can be sectorized as follows:

- Securing the data and content
- Securing the routing
- Authentication and access control
- Certification

In general, solutions need to have certain fixed points that are used to build trust toward the system. This is typically achieved by using some sort of logically centralized identity management. It is known that without this kind of centralized management, the distributed system cannot defend itself from malicious nodes. This is exemplified by the Sybil attack discussed later in this chapter. If attackers may be able to achieve a quorum, a way to control the creation of node identifiers is needed. One potential solution is secure node identifiers through public key cryptography. Also in this case, a mechanism to bootstrap trust toward the public keys is needed.

9.2   Attacks and Threats

Some types of malicious attacks are widely published and warned against; some are less well known and therefore potentially more dangerous. In the following, we briefly discuss typical security threats.

9.2.1   Worms

A worm is a self-replicating software module, designed to use the network for propagation automatically. A worm is an autonomous entity and does not need a carrier code to exist. Worms can carry a payload code of their own and often do so. The payload might be purely destructive, but it can often facilitate secondary malware attacks: launch extortion attacks, send documents via e-mail, install hidden backdoors to create zombie nets (botnets) for illegal mass operations such as spamming and DoS attacks. Worms propagate by exploiting hidden vulnerabilities of operating systems. Worms commonly use e-mail in the form of attachments and Web sites in the form of browser vulnerabilities to compromise end users' machines. Worms propagating through P2P applications would be disastrous: it is probably the most serious threat.

9.2.2   Sybil Attack

In a Sybil attack, the attacker tries to subvert the reputation system in a P2P network [115]. The idea behind this attack is that a single malicious entity can present multiple pseudonymous identities. An entity can have many identities mapped to it for resource sharing and redundancy reasons. Thus the malicious entity can overwhelm the network and gain control over part of the network. After the network has been compromised, the attacker can, for example, gain responsibility over certain files or pollute them.

If the attacking entity can position his identities in a strategic way, the damage from a Sybil attack can be considerable. The attacker might choose to continue in an eclipse attack (presented next), or slow down the network by rerouting all queries in a wrong direction; the adversary can also monitor communications. The vulnerability of a P2P reputation system depends on how easily identities can be generated and how the system reacts to entities without a proper chain of trust.

9.2.3   Eclipse Attack

Overlay networks are vulnerable to *eclipse attacks*, in which malicious nodes in collusion grab the neighboring true nodes [293, 294]. Figure 9.1 illustrates this form of attack. In an overlay network each node maintains links to a set of neighboring nodes (an overlay graph),

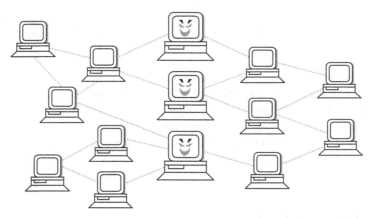

Eclipse attack: the malicious nodes have separated
the network in two subnetworks

FIGURE 9.1
Eclipse attacks in P2P systems.

used to maintain the overlay and to implement application functionality. If a hijacker gets control over a large fraction of the neighbors, it can eclipse the correct nodes and prevent the correct overlay operations. Thus a set of malicious nodes will act in collusion and force the correct nodes to peer only with the corrupted set. The attacker can also utilize the overlay maintenance algorithm and launch a full-size attack from a smaller hijacked set of nodes.

> With eclipse attacks, the adversaries can disrupt overlay communication by taking control over a large fraction of the neighbors of correct nodes even while they control only a part of overlay nodes. Thus mitigation techniques are needed.

The general defense is based on a simple observation that in an eclipse attack the in-degrees of attacking nodes in the overlay graph are necessarily considerably higher than the average in-degree of uncorrupted nodes in the overlay. Therefore, one way to prevent an eclipse attack is for proper nodes to choose neighbors where the in-degree is not significantly above average among the set of nodes satisfying the structural constraints assigned by the overlay protocol.

It may not be sufficient to bound the node in-degrees. Malicious nodes could deplete all in-degrees of correct nodes, thus making it difficult for correct nodes to pair as neighbors. It follows that binding the out-degree of nodes is necessary as well. Proper nodes will then choose neighbors with both in-degree and out-degree below a given threshold.

An eclipse attack mitigation technique has been proposed that uses anonymous auditing to bound the degree of overlay nodes. This technique can be used in homogeneous structured overlays with moderate churn, and it allows important optimizations such as proximity neighbor selection. Experimental results indicate that the technique can prevent attacks effectively in a structured overlay.

9.2.4 File Poisoning

> File poisoning attacks have become commonplace in P2P networks. The goal of this type of attack is to swap a legal file in the network with one provided by the attacker. There are two types of attacks: the content pollution attack and the index poisoning attack [77].

In a content pollution attack, the adversary corrupts the targeted content files that are available for sharing. Because many systems do not test the file for possible corruption, the polluted content rapidly spreads through the P2P network. This type of attack requires substantial bandwidth and server resources to be successful [77].

The second type of attack is the index poisoning attack, in which the adversary inserts large numbers of fake records into the P2P index. This index might be centralized, or it might be distributed over P2P nodes. The most common way to poison records is to use randomized content identifiers instead of the correct ones. Random hash identifiers do not correspond to any existing content in a P2P system, and thus the system cannot locate the content. The extent of damage caused by this kind of attack depends on the size of the corrupted index. The attack can be applied for both structured and unstructured systems.

9.2.5 Man-in-the-Middle Attack

In a *man-in-the-middle (MiTM)* attack, the adversary sits on the communication path of two nodes and, using independent connections to the nodes, relays messages between them. The adversary aims then to be undetected and eavesdrop and manipulate the relayed messages. Man-in-the-middle attacks are a threat in many protocols. As a mitigation strategy, many protocols include some form of end point authentication to prevent this attack.

9.2.6 DoS Attack

A *denial-of-service (DoS)* attack is an attack against a node or a network that involves the creation of vast amounts of connections or packets against the target. The idea is to overwhelm the target and thus disrupt the service or client. The five basic types of DoS are

- Excess consumption of computational resources, such as processor time or bandwidth
- Destruction of configuration information, such as routing tables
- Disruption of state information—e.g., TCP session states
- Destruction of networking hardware—e.g., making fake updates on firmware
- Obstruction of the communication media in the vicinity of the target

In the case of P2P networks, the most common form of a DoS attack is an attempt to saturate the network with packets, thus preventing legitimate network traffic. Another method is to send a vast amount of queries to the target, requiring all of its computational capacity. Because of the large computing power needed, DoS attacks will be more efficient if multiple hosts are used by the adversary. The attack is then called a *distributed denial-of-service (DDoS attack)*. Figure 9.2 illustrates the DDoS attack in P2P systems.

In a DDoS attack, the attacking computers are often personal computers with broadband connections that have been compromised by a virus or Trojan; often they are zombies belonging to a botnet. The attacker is able to remotely control these PCs by directing an attack at any host or network. Furthermore, a DDoS attack can be strengthened by using uncompromised hosts as amplifiers. The zombies send requests to the uncompromised hosts and spoof the zombie IP addresses to the victim's IP. The uncompromised hosts will respond by sending their answering packets to the victim. This is known as a *reflection attack*.

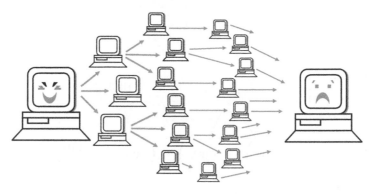

FIGURE 9.2
Denial of service attacks in P2P systems.

9.3 Securing Data

The main goals of content protection in overlay networks are

- Ensuring the availability of content. Preventing deletion and concealment of content.
- Preventing the modification of content.
- Protecting the identity of content publishers and thus enabling publisher anonymity.

The ultimate aim is to make content publishing on networks as secure as possible against various attacks. Many factions such as criminals, hackers, political entities, and even business interests may have interests to suppress or alter content material on the Internet.

There are a multitude of methods to provide content protection or censorship resistance. They are often somewhat overlapping, and certainly many are used in combination to create stronger security tools.

- Self-certifying data
- Merkle trees
- Information dispersal
- Shamir's secret sharing scheme
- Distributed steganographic file systems
- Erasure coding
- Smartcards for bootstrapping trust

In this chapter, solutions for anonymity are considered after examining the above techniques.

9.3.1 Self-Certifying Data

Data is self-certifying when its integrity can be verified by the node retrieving it [227]. A node needing to insert a file in the network will calculate a cryptographic hash of the content of the file to produce the file key. The hash is based on a known hashing function.

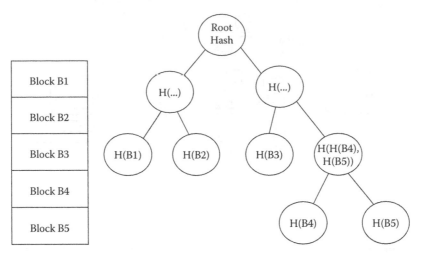

FIGURE 9.3
Example Merkle tree.

Conversely, when the file is retrieved by a node using its key, the node uses the same hash function to verify the data. We should note that the method requires that all nodes in the network share the same knowledge of the hashing function. Similarly, the method requires that the hash-derived signature is used as unique access key for the file. Self-certifying data (and labels) are widely used in P2P and DHT systems.

9.3.2 Merkle Trees

The *Merkle tree* or *hash tree* is a data structure that contains a verifiable summary of a block of data [229]. Merkle trees are commonly used to verify that the content of a given file has not changed with respect to the information contained in the Merkle tree.

The structure is typically based on a binary tree, in which the original data blocks are the leaves, and each level in the tree performs a hash to the digest of the level below. This recursive process results in a single hash value at the root of the tree. This is called the top hash. Figure 9.3 presents an example Merkle tree. The top hash uniquely identifies the file. The hashes are performed with a cryptographic hash function, such as SHA-1. The approach generalizes to trees that have more than two child nodes.

Merkle trees are useful for P2P applications that deal with immutable data because the signatures can be generated based on the data and then stored separately. P2P clients can obtain or verify the root hash of the data from some trusted source. Once this has been done, the hash tree and the pieces of the data can be obtained from nontrusted sources. First any parts of the hash tree received from nontrusted sites are verified against the trusted top hash. Then the actual data can be transferred and verified against the hash tree. One additional benefit of the hash tree is that one branch of the tree can be downloaded at a time and each branch can be checked immediately. This is useful especially with large data sets.

Merkle trees are widely used in distributed systems—for example, in the Sun's ZFS file system and Google's recent Wave service protocol.[1] A variant of Merkle trees based on

---

[1] wave.google.com

the Tiger hash is called Tiger tree hash, and it is widely used in P2P protocols such as Gnutella.

9.3.3 Information Dispersal

> Information dispersal uses an algorithm to split a file into pieces. The file can then be reconstructed from predefined subsets of pieces. Dispersal methods are needed both for fault-tolerant routing and efficient memory management.

Rabin's information dispersal algorithm is in wide use [265]. This algorithm breaks a file F of length L into n pieces, each of length L/m, so that every m pieces are sufficient to reconstruct F. File dispersal and reconstruction are computationally efficient. Rabin's algorithm has numerous applications in secure and reliable storage of information in computer networks and in fault-tolerant and efficient transmission of information in networks.

This algorithm is employed by systems such as Publius [338] and Mnemosyne [159] for encoding information to be published in the network.

9.3.4 Secret-sharing Schemes

Secret sharing is used to designate a method in which a secret is share-wise distributed among the participants. The secret can be reconstructed from its parts, but all shares must be combined together. Generally, in a secret-sharing scheme a dealer gives a secret to n players in such a way that any set of t players (t is the threshold value, $t < n$) can reconstruct the secret in collusion but no set of players smaller than k can perform the reconstruction. If such a method purports to be information theoretically secure, some limitations must be observed:

- A share must be as large or larger than the secret itself. This is a basic requirement stemming from information theory.
- A sharing scheme must use random bits and, in order to distribute a secret of length L, $(t-1) \times L$ random bits are needed, where t is the threshold.

Shamir's secret-sharing system is based on the idea that a unique polynomial of degree $(t-1)$ can be fitted to any set of t points that lie on the polynomial [290]. In Shamir's scheme, a polynomial of degree $t-1$ is created with the secret as the first coefficient and the other coefficients randomly generated. Then, n points are taken on the curve and shared with the players, one to each player. It follows that t points are enough to fit a $t-1$ degree polynomial where the secret is the first coefficient.

In practice, the publisher of the content encrypts a file with a key K, then uses the polynomial method to divide K into n shares so that any k of them can reproduce K, but $k-1$ will give no hints about K. Each participating server encrypts one of the key shares and attaches it with the file. The file becomes inaccessible only if at least $(n-k-1)$ servers containing the key are shut down.

Shamir's secret-sharing scheme is used in several systems (Publius [338] and PAST [116, 278]).

9.3.5 Smartcards for Bootstrapping Trust

We take the PAST system as an example of the usage of smartcards in a large-scale P2P global storage utility [116, 278, 279]. Its native security mechanism relies on the use of smartcards. Each PAST node and each user of the system holds a smartcard. A private/public key pair is associated with each card. Each smartcard's public key is certified by the smartcard issuer's private key. The smartcards generate and verify various certificates used during insert and reclaim operations, and they maintain storage quotas.

A smartcard provides the nodeId for an associated PAST node based on a cryptographic hash of the public key of the smartcard. The assignment of nodeIds probabilistically ensures uniform coverage of the space of nodeIds and also diversity of nodes with adjacent nodeIds—e.g., in terms of geographic location, network attachment, ownership. Furthermore, nodes verify the authenticity of each other's nodeIds.

The smartcard of a user planning to insert a file into PAST issues a file certificate. This certificate contains a cryptographic hash of the contents of the file (computed by the client node), the fileId (computed by the smartcard), the replication factor, and the salt. It is signed by the smartcard. During an insert operation, the file certificate allows each storing node to verify several things:

- The user is authorized to insert the file into the system. This prevents clients from exceeding their storage quotas

- The contents of the file arriving at the storing node have not been damaged en route from the client by faulty or malicious nodes

- The fileId is authentic. This defeats DoS attacks where malicious nodes try to exhaust storage at a subset of PAST nodes using fileIds with nearby values.

Each node that has successfully stored a copy of the file then issues and returns a receipt to the client node. This allows the client to verify that k copies of the file have been created on nodes with adjacent nodeIds, preventing a malicious node from suppressing the creation of k diverse replicas. During a retrieve operation, the file certificate is returned with the file. This allows the client to verify that the contents are authentic.

Before issuing a reclaim operation, the user's smartcard generates a reclaim certificate. The certificate, containing the fileId, is signed by the smartcard. It is included in the reclaim request that is routed to the nodes storing the file. The smartcard of a storage node first verifies that the signature in the reclaim certificate matches the signature in the file certificate. This prevents other than the owner of the file from reclaiming the file. If the operation is accepted, the smartcard of the storage node generates a reclaim receipt. This receipt contains the reclaim certificate and the size of storage reclaimed; it is signed by the smartcard and returned to the client.

9.3.6 Distributed Steganographic File Systems

> Steganography denotes a system which both hides and encrypts information while preventing outsiders from knowing how many files have been stored, if any. The hiding of both existence and number of files gives a measure of plausible deniability [10].

In a distributed steganographic file system, an entire partition is randomized and the encrypted files are hidden within it. Efficient encryption normally generates data that resembles random data; thus, the files will be indistinguishable from the randomized substrate. File locations are coded into the keys for the files, and therefore they will be hidden and available only for legal users. The presence of the files is difficult if not impossible to detect. The system is prepared by first filling all blocks with random data; then files are stored by encrypting their file blocks and placing them at pseudo-randomly chosen locations (normally by hashing the block number with a randomly chosen key). To avoid collisions, a considerable amount of replication will be required.

Generally, the steganographic system requires a large overhead in read/write performance and is inefficient in storage. Still, the plausible deniability is a valuable property in many situations, often more than justifying the complexities of the system [159].

9.3.7 Erasure Coding

> Erasure coding is a method in which a message of n blocks is transformed into a message of m blocks ($m > n$) in such a way that the original message is recoverable from a subset of the m blocks. The fraction of M blocks needed is the rate r of the erasure coding. A mechanism using erasure coding for data durability and a Byzantine agreement protocol for achieving consistency and updating serialization is implemented by the OceanStore system [191].

With erasure coding the data is broken in blocks and spread over multiple servers. Only a fraction of them is needed for regenerating the original block. The objects and the attached fragments are then named. This is done by applying a secure hash to the object contents. This gives them globally unique identifiers and also provides data integrity by ensuring that a recovered file has not been corrupted. A corrupted file would produce a different identifier. Blocks are carefully dispersed, avoiding possible correlated failures. The nodes are chosen to reside in distinct geographic locations or administrative domains.

OceanStore is a global data store designed to be persistent and scalable up to billions of users. It provides a consistent and durable storage utility built atop an infrastructure of untrusted servers. Any computer can join the system and contribute storage or provide local user access. Users need only subscribe to a single OceanStore service provider, although they may use storage and bandwidth from many different providers. The providers automatically buy and sell capacity and coverage among themselves, transparently to the users. In OceanStore, any server may create a local replica of any data object. These local replicas provide access speed and robustness to network partitions. They also reduce network congestion by making access traffic more localized.

Any infrastructure server could crash, become compromised, or leak information. The caching system therefore requires redundancy and cryptographic techniques to protect the data from the underlying servers. OceanStore employs a Byzantine-fault-tolerant commit protocol that provides consistency across replicas. The OceanStore API also permits applications to weaken their consistency restrictions in order to achieve higher performance and availability.

Each object is attached to an inner ring of servers to provide versioning capabilities. The ring maintains a mapping from the original identifier of the object to the identifier of the most recent version of the object. The ring will be kept consistent through a Byzantine agreement protocol [194], allowing $3k + 1$ servers to reach an agreement when no more than k are faulty. Therefore, the mapping is fault tolerant from the active identifier to the most recent identifier. The inner ring also handles the verifying of the legitimate writers of the object and maintains a history of the object updates. Thus it provides, in addition to referential integrity, a universal undo mechanism by storing previous versions of objects.

9.3.8 Censorship Resistance

Publius is one of the tools aiming to cater to the growing problems of Internet censorship and attempts to prevent free flow of content through the network's distribution channels [338]. Publius is a Web publishing system designed to be resistant to censorship while providing publishers with a high degree of anonymity as well. This is important because typically censorship and attempts to break anonymity are simultaneous factors in recent attacks against freedom of information and thus they should be met with a single tool. Aside from censorship, there are also other considerations behind Publius: there is often need to publish content without explicit association with publisher's gender, race, nationality, or ethnic background.

The Publius system consists of the following parts:

- Publishers posting content to the Web
- Servers hosting random-looking content
- Retrievers which browse Publius content on the Web

The Publius system supports any static content and requires a static, system-wide list of available servers. Content is encrypted by the publisher and distributed over some of the Web servers (in the current system, the set of servers is static). The publisher takes the key, K, originally used to encrypt the file and divides it into n parts (shares) such that any k of them can regenerate the original K, but $k-1$ shares are not enough to give any information of the key. Each server receives the encrypted Publius content and one of the shares. At this point, the server has no idea of the hosted content; it simply stores some data that looks random. To browse content, a retriever must get the encrypted Publius content from a server and k of the shares.

The system includes mechanisms for detection of any malicious tampering of the content. The publishing process creates a special URL used to recover the data and the shares; the Publius content is tied to this URL cryptographically, so that any modification of the content or the URL has the result that the retriever is unable to find the information or fails at verification. In addition to the basic publishing mechanism, Publius also provides a way for publishers, and only publishers, to update or delete their own Publius content. Furthermore, it is possible to publish several files simultaneously and also mutually hyperlinked material.

Publius increases system robustness and protects anonymity by distributing redundant partial shares (instead of files) among multiple holders. However, Publius remains imperfect, because the identity of the holders is not anonymized and an adversary could still destroy information by attacking a sufficient number of shares. No publishing system fully protects the consumers of information, although the Rewebber system also operates a separate browser proxy service [147]

9.4 Security Issues in P2P Networks

9.4.1 Overview

P2P DHT-based overlay systems are susceptible to security breaches from malicious peer attacks. One of the simplest attacks on DHT-based overlay system happens when the malicious peer returns wrong data objects to the lookup queries. The authenticity of the data objects can be guaranteed by using cryptographic techniques, either through cost-effective public keys and/or content hashes, securely linking together different pieces of data objects. However, such techniques cannot prevent undesirable data objects from polluting the search results or prevent DoS attacks. Still, malicious adversaries may succeed in corrupting, denying access or responding to lookup queries of replicated data objects. They might also impersonate others so that replicas would be stored on illegitimate peers.

Sit and Morris [296] provide a very clear description of security considerations that involve the adversaries that are peers in the DHT overlay lookup system that do not care to follow the protocol correctly:

- Malicious peers are able to eavesdrop on the communication between other nodes.
- A malicious peer can only receive data objects addressed to its IP address, and thus IP address can be a weak form of peer identity.
- Malicious peers can collude together, giving believable false information.

They also present a taxonomy of possible attacks such as

- Routing deficiencies due to corrupted lookup routing and updates
- Vulnerability to partitioning and virtualization into incorrect networks when new peers join and contact malicious peers
- Lookup and storage attacks
- Inconsistent behavior of peers
- DoS attacks preventing access by overloading network connection
- Unsolicited responses to a lookup query

Securing the content in overlay networks is not sufficient, and also the routing and forwarding processes need to be made secure. The aim of secure routing is to address the problem of malicious nodes attempting to corrupt or delete files, deny access to objects, or poison files and indexes. Therefore, the following processes need to be considered:

- Secure assignment of node identifiers
- Secure maintenance of routing tables
- Secure forwarding of both control and content messages

The solutions used must make these processes secure, naturally combined with other security techniques. Design principles for defenses against such attacks can be classified as follows:

- Defining verifiable system invariants for lookup queries
- Node identifier (nodeId) assignment
- Peer selection in routing
- Cross-checking information by using random queries
- Avoiding single point of responsibility

Castro et al. [62] consider the vulnerabilities of secure routing for structured P2P overlay networks mainly in terms of the possibility that a small number of malicious and conspiring peers could compromise the overlay system. They presented a design and analysis of techniques for secure peer joining, routing table maintenance, and robust message forwarding in the presence of malicious peers in structured P2P overlays. This technique can tolerate up to 25% of malicious peers while providing good performance when the number of compromised peers is small. However, this defense restricts the flexibility necessary to implement optimizations such as proximity neighbor selection and only works in structured P2P overlay networks.

Singh et al. [293, 294] propose a defense that prevents eclipse attacks for both structured and unstructured P2P overlay networks by bounding degree of overlay peers. That is, the in-degree of malicious overlay peers is likely to be higher than the average in-degree of legitimate peers, and legitimate peers thus choose their neighbors from a subset of overlay peers whose in-degree is below a threshold. However, even after the in-degree bounding, it is still possible for the adversary to consume the in-degree of legitimate peers and prevent other legitimate peers from referencing to them. Therefore, bounding the out-degree is necessary; then legitimate peers choose neighbors from the subset of overlay peers with both in-degree and out-degree below some threshold. An auditing scheme is also introduced to prevent incorrect information concerning the in-degree and out-degree.

9.4.2   Insider Attacks

A central problem has been to how find join and leave operations that run as efficiently as possible while still maintaining a highly scalable overlay network. However, besides scalability, robustness is also important, since in open environments like the Internet adversaries may use both insider and outsider attacks on a distributed system [15, 293].

Join and leave operations may be used by attackers to cause trouble in the system. More specifically, we consider the scenario in which there are n honest peers and ϵn adversarial peers in the system for some constant $\epsilon < 1$. The adversary has full control over its adversarial peers and knows the entire overlay network at all points in time. It can use this information to decide in an arbitrary and adaptive manner which of its adversarial peers should leave the system and join it again from scratch. In this way, it is possible to construct sequences of rejoin activities by the adversarial peers in order to harm the overlay network as much as possible—for example, by degrading its scalability or isolating honest peers.

There exists some accumulated experience on join-leave attacks by adversarial peers. The P2P community has been aware of the danger of these attacks, and solutions have been proposed to mitigate the attacks in practice. Until recently no mechanism was known that can be proved to cope with the attacks without sacrificing the essential openness of the system.

The first mechanism shown to preserve randomness for a polynomial number of adversarial rejoin requests uses randomized peer identifiers. A limited lifetime on peers in the system is enforced; i.e., every peer has to reinject itself after a certain amount of time steps. However, this leaves the system in a mode that may unnecessarily deplete resources. The application of competitive strategies would be an ideal solution, and the resources taken by the mixing mechanism should be scalable with the join-leave processes of the system. A corresponding strategy was first presented for a pebble shuffling game on a ring. However, the join rule proposed cannot be directly applied to overlay networks based on the concept of virtual space. This is because it has no control over the distribution of the peers in the virtual space. The balancing condition might be violated.

The first rule able to satisfy both the balancing and majority conditions while accepting a polynomial number of rejoin requests from attacking peers is the k-cuckoo rule, outlined in Figure 9.4. The rule requires that a new peer pick a random identifier x and then all peers in the x's k-region of size k/n are relocated to points in the identifier space that are chosen uniformly and independently at random. This rule is a randomized join strategy that wins with high probability against any adversary as long as $\epsilon < 1 - 1/k$. The aim of the join strategy is to maintain two conditions—namely, that in any $\Theta((\log n)/n)$ interval there are $\Theta(\log n)$ nodes, and that the good nodes are in the majority in every such interval. An improved version of the rule called cuckoo and flip rule has been proposed that alleviates region-balancing issues when nodes leave [15].

9.4.3   Outsider Attacks

Besides insider attacks, we must also consider certain outsider attacks. The adversary might shut down any peer at any point in time by starting a brute-force DoS attack bypassing the overlay network. We assume that an honest peer that is exposed to such an attack will leave the system and will rejoin the network anew as soon as the DoS attack on it is over.

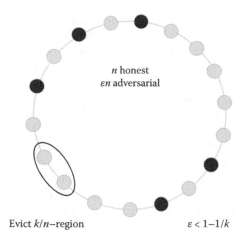

Evict k/n–region $\varepsilon < 1-1/k$

Rejoin: leave and join via k-cuckoo rule

FIGURE 9.4
k-cuckoo rule.

There are two types of serious forms of attack:

1. Consider outsider attacks that are oblivious and targeted on P2P systems with peers placed at randomized positions in a virtual space. These can be modeled as random faults or churn. Random faults and churn has been heavily investigated in the P2P community, and it is known that some structured P2P systems like Chord can tolerate any constant probability of failure.

2. Adaptive outsider attacks are more difficult to handle. The best current method results in a structured overlay network that can recover from any sequence of adaptive outsider attacks in which at most $\log n$ peers may be removed from the system at any point in time. The basic idea behind this approach is that the peers perform load balancing locally in order to fill the holes the adversary may have caused in the network. This approach works fairly well when peers in the network can be assumed to be honest, but it fails if some of the peers are adversarial. The adversary would only have to focus on a particular corner of the network and force all honest peers in it to leave until enough adversarial peers have accumulated in the corner. If the adversarial peers are able to gain the majority in this corner, they can launch serious application-layer attacks. It will not be possible any more to wash out adversarial behavior in a proactive manner.

9.4.4 SybilGuard

Decentralized, distributed systems like P2P systems are particularly vulnerable to Sybil attacks. Sybil attack (termed by Douceur [115]) describes the situation whereby there are a large number of potentially malicious peers in the system without a central authority to certify peer identities. In a Sybil attack, a malicious user obtains multiple fake identities pretending to be multiple, distinct nodes in the system. Control over a large fraction of the nodes in the system gives the malicious user ability to throw out the honest users in collaborative tasks such as Byzantine failure defenses.

In such a situation, it becomes very difficult to trust the claimed identity. Dingledine et al. propose puzzle schemes, including the use of microcash, which allow peers to build up reputations [113]. Although this proposal provides a degree of accountability, it still allows a resourceful adversary to launch successful attacks. Many P2P computational models of trust and reputation systems have emerged to assess trustworthiness behavior through feedback and interaction mechanisms. These computational trust and reputation models make the basic assumption that the peers commit themselves to bilateral interactions and evaluations on a globally agreed scale.

SybilGuard is a protocol that limits the corruptive influences of Sybil attacks. The protocol is based on the social network of user identities. An edge between two identities points out a human-established trust relationship. Malicious users can create multiple identities, but in general they are unable to forge many trust relationships. Thus, the cut in the graph between the sybil nodes and the honest nodes is too small. SybilGuard exploits this property for setting an upper bound for the number of identities of a malicious user. The effectiveness of SybilGuard has been shown both analytically and experimentally [355].

9.4.5   Reputation Management with EigenTrust

The trustworthiness of a node can be measured by the level of trust that other peers have toward the node. This activity of collecting reputation information in a distributed environment, and then making decisions based on it, is called reputation management. Reputation management involves a number of processes, including the following:

- An entity's actions are tracked.
- Other entities' opinions about those actions are tracked.
- Actions and opinions are reported.
- Reacting to the report is enhanced by creating a feedback loop.

Reputation management uses preset criteria and algorithms for processing complex data for tracking and reporting reputations. These systems facilitate the process of determining trustworthiness automatically.

> EigenTrust is an algorithm for reputation management in P2P networks, developed by Sepandar Kamvar, Mario Schlosser, and Hector Garci-Molina. The algorithm calculates a unique global trust value for each peer in the network. The value is based on the upload history of the peer and helps to cut down the number of incorrect files in a P2P file-sharing network [177].

The EigenTrust algorithm makes use of the concept of transitive trust: If a peer i trusts any peer j, it would trust peers trusted by j as well. Each peer i evaluates the local trust value s_{ij} for all peers having provided authentic or bogus downloads based on the satisfaction level of its experienced transactions. More formally,

$$s_{ij} = sat(i, j) - unsat(i, j), \tag{9.1}$$

where $sat(i, j)$ refers to the number of adequate responses that peer i has received from peer j and $unsat(i, j)$ refers to the number of inadequate responses that peer i has received from peer j. The local value is always normalized, thus preventing malicious peers from assigning arbitrarily high local trust values to their colluding malicious partners and, respectively, arbitrarily low local trust values to satisfactory peers. The normalized local trust value c_{ij} is given by

$$c_{ij} = \frac{max(s_{ij}, 0)}{\sum_j max(s_{ij}, 0)}. \tag{9.2}$$

The system uses either the aggregation of the local trust values at a central location or accesses them in a distributed manner in order to create a trust vector for the whole network. According to the concept of transitive trust, a peer i would ask any other peers it knows to report the trust value of a peer k, weighing the responses by the trust peer i places in them. This trust value is given by t_{ik}:

$$t_{ik} = \sum_j c_{ij} c_{jk}. \tag{9.3}$$

Assuming that a user knows the c_{ij} values for the network, given by the matrix C, then trust vector $\bar{t}_i$ defining the trust value for t_{ik} is given by

$$\bar{t}_i = C^T \bar{c}_i. \tag{9.4}$$

In the equation above, if C is aperiodic and strictly connected, powers of the matrix C will be converging to a stable value at some point.

$$\bar{t} = (C^T)^x \bar{c}_i. \tag{9.5}$$

For large values of x, the trust vector $\bar{t}_i$ appears to converge to a single vector for every peer in the network. The vector $\bar{t}_i$ is the left principal eigenvector of the matrix C. It is notable that, since $\bar{t}_i$ is the same for all nodes in the network, it stands for the global trust value.

It is possible to develop a simple centralized algorithm computing trust value. We assume that all local trust values for the network are available and presented in the matrix C. If the equation is convergent, the initial vector $\bar{c}_i$ can be replaced with a vector $\bar{e}$ that is an m-vector and represents the uniform probability distribution over all m peers. Algorithm 9.1 presents a simple centralized version of the EigenTrust algorithm.

Algorithm 9.2 presents the basic EigenTrust algorithm that addresses some limitations in the simple algorithm. The simple algorithm does not address a priori notions of trust, inactive peers, and malicious collectives. Pretrusted peers are crucial for the basic algorithm, because they guarantee convergence and break malicious collectives. The choice of the pretrusted peers is important, and that they are not members of malicious collectives. The pretrusted peers are represented by $\bar{p}$, which is some distribution over them. In the algorithm a is a system parameter, a constant less than 1.

The Eigentrust system has also been extended for decentralized environments, in which a component-wise algorithm is used. In this case, each peer stores its local trust vector and trust value. Security is provided by assigning trust score, managed using a DHT such as CAN or Chord [177].

Algorithm 9.1 A simple centralized EigenTrust algorithm

Function: *SimpleEigenTrust()*
$\bar{t}^0 = \bar{e}$
repeat
$\quad \Big| \quad \bar{t}^{(k+1)} = C^T \bar{t}^{(k)}$
$\quad \Big| \quad \delta = ||t^{(k+1)} - t^{(k)}||$
until $\delta <$ error

Algorithm 9.2 The basic EigenTrust algorithm

Function: *EigenTrust()*
$\vec{t}^0 = \vec{p}$
repeat
$\quad$ $\vec{t}^{(k+1)} = C^T \vec{t}^{(k)}$
$\quad$ $\vec{t}^{(k-1)} = (1+a)\vec{t}^{(k+1)} + a\,\vec{p}$
$\quad$ $\delta = ||t^{(k+1)} - t^{(k)}||$
until $\delta <$ error

9.5 Anonymous Routing

One of the central requirements for P2P networks is anonymity. Anonymity is often accomplished by employing anonymous connection layers, such as onion routing [112, 148] and mix networks [3, 69]. For instance, the anonymizing, censorship-resistant system proposed in [288] splits documents into encrypted shares using an anonymizing layer of nodes (referred as forwarders). The technique then chooses nodes that store the shares and destroys the original data. Data requests and resulting shares are forwarded by using onion routing thus anonymizing the addresses. FreeHaven [114] is a similar system built on top of anonymous remailers to provide necessary pseudonyms and communication channels.

This section describes the main approaches currently used for providing anonymity to routing processes in overlay networks, especially in P2P systems. The Freenet system was already presented in Chapter 4, and in the rest of this section we present additional examples of systems that support anonymity.

9.5.1 Mixes

Mixes introduced the notion of anonymous digital communication, and the Mix system essentially creates unlinkability between sender and receiver [69]. This ensures that, while an attacker can determine that the sender and receiver are actually communicating by sending and/or receiving messages, he cannot detect with whom they are communicating. The system is composed of a mixed set of nodes that store, mix, and forward the messages in transit. The route of the message is predetermined by the sender using one or more mix nodes and a precisely defined protocol. A public key cryptography protocol is used to ensure that messages cannot be tracked by an adversary while passing through the mix network. In the simplest form that is called a *threshold mix*, a node waits until it is able to collect a set of messages as input. The private key is then used to get the address of the next mix node or final destination. The received and buffered messages are then reordered according to some metric before forwarding them. In this sense, an attacker cannot trace a message from source to destination without the help of the mix nodes.

Kesdogan et al. provide a mix-network routing protocol by introducing the *free route* and *mix cascade* concepts [185]. Free route gives autonomy to the sender to choose dynamically the trust path for the mix nodes, but in mix cascade the routing paths are preset. Mix networks use delays due to buffering and mixing and different padding patterns for mixing real and dummy traffic. Continuous mixes try to avoid delay issues by introducing fixed delay distributions. Mixes have been subject to several attacks, such as timing attacks, statistical message distributions analyses, and statistical analysis of the properties of randomized routes.

9.5.2 Onion Routing

Onion routing [148] is an overlay infrastructure created to provide anonymous communication over a public network. Onion routing supports anonymous connections in three separate phases: connection setup, data exchange, and connection termination.

In the setup phase, the initiator will create a layered data structure called onion that implicitly defines the complete route through the network. An onion is encrypted recursively by applying public key cryptography. The number of encryptions is equal to the number of *onion routers (OR)* that process the onion while it is moving toward the destination. The outer cryptographic control layer pertains to the first router in the onion path, while the innermost cryptographic control block refers to the last onion router in the path (i.e., the predecessor to the destination). Each router along the onion route uses its public key to decrypt the next layer off the onion that it receives. This operation reveals the embedded onion, and also the identity of the next onion router. Essentially, each onion router takes off one layer of encryption to arrive in plain form at the next recipient.

Each onion router also pads the embedded onion after decrypting a layer to maintain a fixed size, and then sends it to the next router in the onion chain. Once the onion reaches the destination, all of the inner control data appears as plain text. Essentially this establishes an anonymous end-to-end connection, and data can be sent in both directions. For data moving backward through the connection, the layering occurs in the reverse order and also applies different algorithms and keys. The tear-down of the connection can be started by either end and also in the middle of the path if need be. All messages (onions and real data) transferred through the onion routing network will be identically sized to the messages arriving at an onion router using fixed time intervals. However, the messages are mixed to avoid correlation efforts by potential attackers and, additionally, cover traffic in the semipermanent connections between onion routers can misguide external eavesdroppers.

9.5.3 Tor

Tor[2] is a circuit-based low-latency anonymous communication service in which the second-generation onion routing system addresses limitations by adding perfect forward secrecy, directory servers, congestion control, integrity checking, configurable exit policies, and an architecture for location-hidden services using rendezvous points [112]. Tor needs no kernel modifications nor special privileges. Furthermore, it requires very little synchronization or coordination between nodes, at the same time providing a reasonably balanced solution between anonymity, usability, and efficiency.

Using Tor protects, for example, against traffic analysis, a common form of Internet surveillance activity. Traffic analysis can be used to detect which nodes are communicating over a public network. Knowing the source and destination of the traffic allows adversaries to track one's behavior and interests. This information could be very useful, for instance for e-commerce sites adjusting prices and conditions based on the country or institution of the shopper. It could even pose a safety threat by revealing the identities and geographical locations, even if the connection is encrypted.

The traffic analysis uses the knowledge that Internet data packets have two parts: a data payload and a header used for routing. The data payload may be encrypted, but the header discloses information such as the source, destination, size, timing. The header part will reveal more than most users would like it to, especially under statistical analysis based on cumulative data.

---

[2] www.torproject.org

A basic problem for Internet privacy is that the recipient of the communications can see who sent it simply by looking at headers, as could authorized intermediaries like service providers, and often unauthorized intermediaries as well. The simplest, but very effective, form of traffic analysis involves a node situated between sender and recipient on the overlay network, just looking at headers. Nevertheless, there are also more powerful methods of traffic analysis. Some adversaries watch on multiple locations of the Internet using sophisticated statistical techniques to track the communications patterns of target organizations and individuals. No encryption helps against these attackers, since it can only hide the content of the traffic, not the header information.

Tor applies the concept of onion routing. Onion routing is a distributed overlay network built to facilitate making anonymous many TCP-based applications such as Web browsing, secure shell, and instant messaging. Clients choose a path through the network and build a circuit in which each node (onion router) in the path knows its predecessor and successor but not the other nodes in the circuit. Traffic moves in the circuit in fixed-size cells, unwrapped by a symmetric key at each node (like peeling an onion) and then sent downstream. However, real-Internet applications are scarce: while a wide-area onion routing network was briefly tested, the only notable public implementation was a single machine proof-of-concept experiment, which was successful in itself but not conclusive. Many critical design and deployment issues were not resolved, and the design has not been updated in years. Nevertheless, the onion concept was deemed useful enough to be implemented in more advanced designs, and Tor is one of those.

The Tor network is an overlay network where each onion router is running as a user-level process without having any special privileges. Each onion router maintains a TLS (transport later security) connection [111] with every other onion router. The users run a local software package called an *onion proxy (OP)* to establish circuits across the network, retrieve directories, and handle connections with user applications. The onion proxies accept TCP streams, multiplexing them across the onion circuits. The onion router at the other end of the circuit takes connection to the requested destination relaying data.

Each onion router maintains two keys: an identity key that is long term and an onion key, which is short term. The identity key is used in three ways: to sign TLS certificates, to sign the descriptor for OR router (this is a summary of the keys, address, bandwidth, exit policy, etc.), and to sign the directories by the directory servers. The onion key is used to decrypt user requests for setting up a circuit and to negotiate ephemeral keys. The TLS protocol is also used to establish a short-term link key if it is communicating between ORs. Short-term keys will be rotated periodically and independently; this is done in order to limit the impact of possibly compromised keys. Figure 9.5 illustrates the building of a two-hop circuit with Tor and using it to fetch a Web page.

Tor also enables users to hide their locations. This is very important when offering services such as Web publishing or an instant messaging server. Rendezvous points help other Tor users to connect to the hidden services without knowing the other participant's network identity. This hidden service functionality allows Tor users to set up Web sites for publishing any material without worrying about censorship. It would be impossible to determine who is offering the site, and nobody who offers the site would know who was posting to it.

9.5.4 P2P Anonymization System

The goal of anonymization of overlay networks is to help a node in communicating with an arbitrary other node in such a way that the identity of the node is impossible to determine. The P2P anynomization system (formerly Tarzan) [139, 140] was proposed to take care of this requirement for anonymity. The Tarzan system is a decentralized network layer

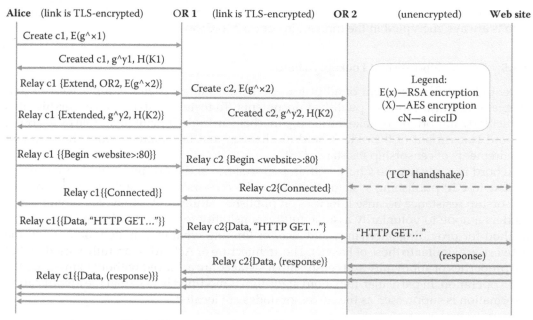

Alice builds a two-hop circuit and begins fetching a Web page

FIGURE 9.5
Overview of Tor.

infrastructure that builds anonymous IP tunnels between an open-ended set of peers. By using the Tarzan infrastructure, a client is able to communicate with a server and nobody can determine the identity of either one.

The anonymizing layer is fully decentralized and is also transparent to both clients and servers. System nodes communicate over sequences of mix relays. These are chosen from an open-ended pool of volunteer nodes, without any centralized component. Tarzan includes techniques that enable secure discovering and selecting of other nodes as communication relays. All peers will be feasible originators of traffic; all peers are potential relays as well. The scalable design greatly lessens the significance of targeted attacks and also inhibits network-edge analysis; a relay is not able to tell when it is the first point in a mix path. The system also works to remove potential adversarial bias: an adversary may run hundreds of virtual machines, but it will be unlikely for it to control hundreds of different IP subnets.

Tarzan also introduces a scalable and useful technique for covering the traffic. This uses a restricted topology for packet routing. Packets can be routed only between two *mimics*. They are pairs of nodes assigned by the system in a way that is secure and universally verifiable. The technique does not expect network synchrony and takes only a little more bandwidth than the original traffic that is to be hidden. The Tarzan technique shields all network participants, not only core routers.

Thus packets are routed in Tarzan via tunnels of randomized peers and using a mix-style layered encryption very similar to onion routing. The two ends of this communication tunnel are a Tarzan node running a client application and a Tarzan node running a network address translator. The translator forwards the traffic to the final destination, an ordinary Internet server. The system is essentially transparent to both client applications and servers. However, it must be installed and configured on all participating nodes. A possible policy that would further reduce the risk of attacks would be for the tunnels to contain peers from different jurisdictions or organizations. Nevertheless, some performance would then be sacrificed. Crowds is a system similar to Tarzan; the core difference is that in Tarzan the data is always encrypted in the tunnels, in contrast to Crowds.

9.5.5   Censorship-resistant Lookup: Achord

A central and fixed element of all peer-to-peer publishing systems is a mechanism that enables efficient locating of published documents. To foster resistance to censorship, it is particularly important to make the lookup mechanism both difficult to disable or to abuse. Achord is a variant of the older Chord mechanism that takes into account the stringent requirements of censorship resistance [161].

Achord is equivalent to Chord both in performance and correctness, but more suitable for use in P2P publishing systems that aim to be censorship resistant. Achord provides censorship resistance because it focuses on publisher, storer, and retriever anonymity and hinders a node to voluntarily assume full responsibility for a certain document. Its basic method for providing anonymity and also for limiting what each node will know of the network are similar to those of Freenet. The architecture of Achord is carefully varied so that the properties of anonymity and censorship resistance are achieved without hampering the main operation. In particular, the Chord algorithm is modified here so that the identification information is suppressed as the successor nodes are located.

9.5.6   Crowds

Crowds is a network that consists of multiple nodes collaborating voluntarily [269]. The basic idea is that the anonymity of a single object can be protected better when it is moving within a crowd of objects. Crowds Web servers are unable to learn the origin of any request because all members of the set of potential requestors are equally likely. Even in collaboration, the Crowds members are not able to distinguish the originator of a request from a member forwarding the request on behalf of some other member. In Crowds, each user is represented in the system by a *jondo process*. When a message that will require user anonymity arrives at the Crowds node, its arrival is announced using the local jondo. Then it is sent to another jondo, randomly chosen, with probability p or to the actual server with probability $1 - p$. When the server (or recipient jondo) receives the message, it will respond using the same forwarding path. Crowds can effectively prevent traceback attacks and also relieve collusion attacks if the users select the set of forwarding jondos by randomization.

9.5.7   Hordes

Hordes [198] is an anonymizing infrastructure combining elements from both onion routing and Crowds. It is the first protocol that used multicast transmission when the destination answers the sender. It includes two phases: the initialization and the transmission phase.

In the initialization phase, Hordes uses the jondo concept from Crowds. In addition, it uses a public key scheme to give authentication services. The sender dispatches a join-request message to a proxy server, which will authenticate the sender by returning a signed message. This includes the multicast address of jondos and tells the multicast group of the new entry.

In the transmission phase of a message, the sender picks a subset of jondos to be used in the forwarding path and also a multicast group address for the reverse path. When a data message becomes scheduled for transmission, the sender will select a jondo member within the forwarding subset, sending the message to this peer as an encrypted onion-type data structure. The designated jondo then sends this message either to another random jondo with probability p or directly to the receiver with probability $1 - p$. Encryption layers are used throughout. The receiver replies using the preset return path. It will also send an acknowledgment to the multicast group as a plaintext message.

9.5.8 Mist

A new system that tackles some of the privacy drawbacks discussed above is the Mist [8]. The Mist handles the problem of routing messages in a network but keeps the location of the sender strictly concealed from all intermediate devices (such as routers and caching tools), from the receiver, and from any potential eavesdroppers. The system is built with a number of routers (Mist routers), which are ordered in a hierarchical structure. Mist also applies special routers, called *portals*, which are enabled to be aware of the user locations, yet without knowing the corresponding identities. The designated *lighthouse routers (LIGs)* are aware of the user's identity, but even they do not know their exact location. The emphasis of the Mist architecture is the distributed knowledge. Due to the decentralized nature of Mist, a possible malicious collusion by some of the Mist routers is nearly impossible because the Mist routers are ignorant of each other's identity. The leaf nodes in the Mist hierarchy (portals) function as points where users are connected to the Mist system.

For example, let us assume that publisher X requires a network service ensuring privacy and data confidentiality. X must first register with the Mist system. The publisher's device interfaces directly with one of the portals available in the nearby Mist space. The portal replies to the request with a list of its ancestral Mist routers that exist at a higher level within the hierarchy and are willing to act as a LIG (point of contact) for the user.

Subscribers intending to have communication with publisher X have to contact his designated LIG. After the LIG selection, a virtual circuit (a Mist circuit) must be newly created between publisher X and the corresponding LIG. This Mist circuit establishment process aims to enable the LIG of publisher X to authenticate X without revealing the physical location of X. Simultaneously, Mist hides the identity of X and the designated LIG from the portal. Furthermore, the Mist circuit applies a routing technique, which is hop-to-hop and handle-based, for packets transmitted between source and destination nodes. Also, in combination with data encryption, it will conceal any information on the identities and locations of the communicating parties.

To establish a Mist circuit, X will generate a circuit establishment packet, transmitting this packet to the corresponding portal. X does not inform the portal of the selected LIG. When it receives the packet, the portal assigns a special handle ID number to the current communication session with X. The portal encloses the assigned handle ID in the received packet and forwards it to its ancestor in the Mist router chain. When the data packet moves through the Mist hierarchy, each LIG router makes an attempt to decrypt the payload by applying their private key. If the decryption fails, the particular router will decide that it cannot be the final recipient of this packet, forwarding the packet to the next router in the hierarchical chain.

The process is repeated by every intermediate Mist router until the packet reaches the ultimate destination. If the decryption of the payload is successfully performed, this forms an indication that X has in fact selected the current Mist router as his LIG. The LIG then responds to X confirming the registration. From there, a secure circuit will be established and X can communicate securely with its LIG. Note that even though the LIG of X can infer that its physical location is underneath a given Mist router Y, it is very hard to determine X's exact position. After the circuit establishment, the LIG will accept the role of representing the end-user.

A further issue to be handled is the detection of the user's LIG. A public directory—for instance, a lightweight directory access protocol (LDAP) server or a plain Web server—may be used for this purpose. For example, subscriber Z tries to communicate with publisher X.

X and Z have previously established Mist circuits with corresponding LIGs LZ and LX. Z transmits to LZ a packet indicating that he wants to set up a publish/subscribe (pub/sub) service with X. LZ will verify that the originator of the message is really Z, locates LX (the LIG of publisher X), and carries out the initialization for establishment of the connection. If the communication path is successfully established, X and Z are able to communicate with each other. Here the intermediate routers are always unaware of the two end points of the communication. Moreover, it is impossible for Z to detect the location of X and vice versa.

9.6   Security Issues in Pub/Sub Networks

Many security concerns emerge in pub/sub overlay environments (because of the many-to-many communication model) with regard to authenticity, integrity, confidentiality, and also availability. For instance, one has to be able

- To guarantee that only authentic publications are delivered to the subscribers (publication authenticity) and that only the subscribers really subscribing to the service will get publications matching their interest (subscription authentication)
- To prevent unauthorized, possibly malicious, modifications of pub/sub messages (publication and subscription integrity)
- To perform the content-based routing without the publishers trusting the pub/sub network (publication confidentiality)
- To not reveal their subscriptions to the network (subscription confidentiality).
- To protect the pub/sub services from any spamming or flooding attacks, both selective and random message dropping attacks, and other DoS attacks.

Many attacks gravely threaten message integrity (unauthorized write) and authenticity (false origin) in addition to confidentiality of messages (unauthorized read). Yet most of the existing secure event-distribution protocols focus only on content confidentiality. Relatively trifling effort has been devoted to developing a more coherent security framework that would be able to guard the pub/sub system from multiple native security problems.

9.6.1   Hermes

Access control is a crucial security requirement, especially in commercial pub/sub applications. Access control is used to assign privileges to all elements participating in the pub/sub architecture. Hermes [255] pub/sub system is a distributed event-based middleware architecture adapted to a type-based and attribute-based publish/subscribe model. Hermes is built around the notion of an event type, and it will support features derived from object-oriented languages such as type hierarchies and supertype subscriptions. A scalable routing algorithm atop of an overlay routing network is used, thus avoiding global broadcasts, since rendezvous nodes are created. Fault-tolerance mechanisms able to cope with breakdowns of the middleware are fully integrated with the routing algorithm. This yields a scalable and robust system.

The main goal of the Hermes architecture is to create a system in which security is managed and controlled within the pub/sub middleware, access control being fully transparent to both publishers and subscribers. In Hermes, each event has a designated owner identified with a X.509 certificate. These owners decide upon the access policies for their own events. Users are then assigned roles, and, furthermore, privileges are assigned to each role.

However, the users will never be assigned privileges directly. This approach has two clear advantages: administration of privileges is much easier, and policy control becomes strictly decoupled from the specific software under protection. Both publishers and subscribers are to be authenticated. Every request they direct to brokers is delivered using their own credentials. Based on these credentials, brokers may then either accept, partially accept, or reject the request. Policies are expressed in a specific policy language provided by OASIS [27].

In Hermes, decisions upon access control are always based on predicates. Generic predicates are used, and they are operated as black boxes. A predicate might make decisions on the basis of the size of the message. The predicates could be publish/subscribe restriction predicates, as well. In that case, predicates are fully understood by the pub/sub system, and they will use the event-type hierarchy. For example, if a subscriber attempts to subscribe to any event that it is not authorized to access, the system will try to detect if the subscriber is authorized to subscribe to any subtypes of the event. Thus the original subscription request is transformed to a different subscription scope. For this approach to be effective, brokers must be trusted to use the access control policies. The Hermes architecture also allows for the usage of certificate chains, forming a web-of-trust. In this web-of-trust, an event owner will sign the trust broker certificates, while these brokers will sign the certificates of their immediate brokers, etc. If publishers and subscribers can show a trusted root certificate for the event owners, they can verify whether their local brokers are eligible to process a certain event.

9.6.2 EventGuard

EventGuard [303] is a mechanism designed to provide access security in content-based pub/sub systems. The goals are to provide authentication for publications, confidentiality and integrity for both publications and subscriptions, and to enable availability while not forgetting performance metrics, scalability, and ease of use.

EventGuard is a modular system designed to operate above a content-based pub/sub core. It has two main components:

- A suite of security guards seamlessly pluggable into a content-based pub/sub system.
- A flexible pub/sub network design capable of scalable routing and handling of message dropping DoS attacks and node failures.

EventGuard uses six guards that secure six critical pub/sub operations (subscribe, advertise, publish, unsubscribe, unadvertise, and routing) as well as a metaservice that generates tokens and keys. Tokens are used as identification of the publications, such as a hash function over publication topic. Keys are used for encrypting message contents. All pub/sub operations involve communication with the provided metaservice before sending any messages. The El-Gamal algorithm is used for encryption, signatures, and the creation of tokens [303].

9.6.3 QUIP

QUIP is a protocol used for securing content distribution in P2P pub/sub overlay networks [89]. It is designed to provide encryption and authentication mechanisms for already existing pub/sub systems. QUIP's security goals are the following:

- To protect content from unauthorized users
- To protect payment methods and to authenticate publishers
- To protect the integrity of the communicated messages

QUIP tackles two main problems—first, ensuring that subscribers are able to authenticate the messages they are receiving from publishers, and second, ensuring that publishers; are able to strictly control who receives their content. QUIP allows the application of a public key traitor-tracing scheme. This scheme has two main advantages:

- The ability to invalidate the keys of any subscribers without hampering the keys of the others.
- Each subscriber has a unique key that facilitates detecting who has leaked a key.

The main idea is to incorporate the efficient traitor-tracing scheme with a secure key management protocol. This allows publishers to restrict their messages to truly authorized subscribers and also to add and remove subscribers without affecting the keys held by the other subscribers.

QUIP does not deal with the problems of privacy in subscriptions. It assumes a single trusted authority fully responsible for keys and payments handling. This is the key server. Each participant in the pub/sub network willing to use QUIP has to download in advance a QUIP client that will provide the participant with a unique random ID and also the public key of the key server. At the initiation, the key server provides a certificate to any QUIP participants linking their public key to their identifier. A publisher wishing to publish a protected object contacts the key server. In return it will receive a content key, which is then used for encryption. Subscribers wanting to read the encrypted publication have to contact the key server in order to obtain the content key (this may require payment, depending on the system).

10

Applications

This chapter considers applications of overlay technology. We focus on four key application areas—a commercial service system, video delivery, P2P session initiation protocol (P2PSIP), and content delivery networks (CDNs). Our commercial service example is the Amazon Dynamo, which uses many techniques from distributed computing and overlay systems, such as vector clocks, gossip, consistent hashing, replication, and reconciliation. Then we discuss video delivery using overlay technologies, with emphasis on video-on-demand (VoD). Video traffic is becoming increasingly popular, and thus techniques are needed to ensure efficient content distribution for both real-time and on-demand media. Then we consider the P2PSIP protocol, which aims to offer a decentralized version of the SIP signaling protocol. Finally, we discuss CDNs based on both commercial and DHT (distributed hash table) technologies.

10.1   Amazon Dynamo

Rapid growth in e-commerce and the proliferation of different types of commercial activities have created a simultaneous demand for new technological solutions, systems, and platforms. The weight of this demand has been in overlay networks. There are many examples of this concurrent development of e-commerce and its tools where the infant market has launched the first versions of technology and a successful system design has helped to mature the commerce, starting a continuous feedback cycle of technological solutions and commercial possibilities. A good example of the process is Amazon, world's largest bookseller, which runs a world-wide e-commerce platform serving tens of millions of customers and using many data centers around the world with tens of thousands of servers.

Amazon Web services stack operates in three levels:

1. E-commerce solutions
   - E-commerce service
   - Historical pricing service
   - Alexa thumbnails
   - Mechanical turk (answers)
2. Search solutions
   - Alexa web search platform
   - Alexa top sites
   - Alexa web information service

3. Infrastructure solutions
   - Messaging (simple queueing service)
   - Storage (simple storage service)
   - Grid (elastic compute cloud)

Understandably, the Amazon operating platform faces stringent requirements for performance, reliability, and efficiency. Also, to support continuous business growth, the platform must be highly scalable. The most important requirement is reliability, because even slight malfunctions can cause financial losses and, furthermore, will impact the customer trust adversely.

The Amazon platform is decentralized, loosely coupled, and service oriented, resulting in an architecture with hundreds of separate services. In the Amazon business environment there is a particular need for storage technologies of high availability. Disks may fail, network routes may be unreliable, or data centers may face electricity problems. Despite all this, customers should be able to view their shopping carts and add items to it. The shopping cart manager service has to be able to write to and read from data stores, and the multiple data centers must always have the data available. Ability to deal with failures in an infrastructure consisting of a very large set of components must be a central tenet of the standard mode of operation. At any given time there are always a significant number of server and network components in danger of failure, and Amazon's software systems see the ability of failure handling without troubling availability or performance as a central requirement.

Amazon relies heavily on relational databases that are the central powering elements of any system of e-commerce where visitors must be able to browse and search for a large number of products. Modern relational database should be able to do this. However, there are problems with redundancy and parallelism in large relational databases, and they might become a point of failure mainly because processing replications is nontrivial. For example, if two database servers need to have identical data, then synchronizing the data will be difficult when both servers are actively reading and writing. The master/slave model is not applicable, because the master will be burdened with the users writing information. Thus, when a relational database grows, its limitations will rapidly become evident and may throttle the entire system. Laterally adding more Web servers does not remedy this situation.

Figure 10.1 illustrates the environment in which Dynamo and other Amazon core services are used. The actual Web pages are created by page-rendering components run on Web servers. A request routing infrastructure is used to connect the page-rendering components, aggregator services, and the data storage services together. Dynamo is used for the storage and maintenance of profile information, such as shopping carts and user preferences.

Amazon has developed a number of storage technologies to attain the requirements for reliability and scaling. The most important of these is probably Dynamo[105], a Java-based, highly available, and scalable distributed data store meant for the Amazon platform, where services typically have very high reliability requirements. The platform requires that the balance between availability, consistency, cost-effectiveness, and performance must be closely controlled. The Amazon commercial platform has a very large and diverse set of applications with a plethora of storage requirements.

Some applications require flexible storage technology where application designers can configure their data store specifically in a cost-effective manner. This will always result in trade-offs to attain high availability simultaneously with guaranteed performance. Still, a primary-key access to a data store is enough for many services on the Amazon platform.

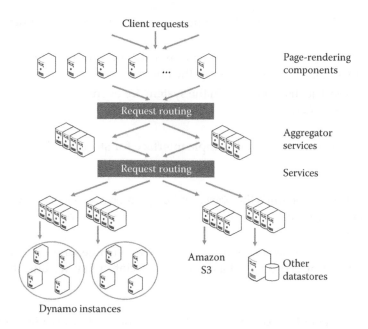

FIGURE 10.1
Amazon Dynamo.

To this group belong the central services of providing shopping carts, customer preferences, best seller lists, session management, product catalogues. Here the normal usage of a relational database would lead to inefficiencies severely limiting both scalability and availability. For these applications, Dynamo provides a primary-key-only interface.

In order to attain scalability and availability, Dynamo uses a combination of techniques:

- Consistent hashing is used to partition and replicate data objects.
- Object versioning is used to increase consistency.
- Quorum-like techniques are used to maintain consistency among replicas during updates.

10.1.1   Architecture

A complex architecture for a storage system is necessary to operate in a production setting. At least the following scalable and robust components are needed: components for load balancing, data persistence, membership detection, failure detection, failure recovery, state transfer, overload handling, replica synchronization, request routing, concurrency and job scheduling, request marshalling, system monitoring, and configuration management. The Amazon Dynamo architecture consists of the following key components and features:

- Nodes: Physical nodes are identical and organized into a ring. Virtual nodes are created by the system and mapped on physical nodes, enabling hardware swapping for maintenance and failure. Any node in the system can receive a put or get request for any key.
- Partitioning: The partitioning algorithm specifies which nodes will store a given object. The mechanism is automatically scaled when nodes enter and leave.

- Replication: Every object will be asynchronously replicated to a number of other nodes.
- Updating: The updating is done asynchronously and thus might result in multiple copies with slightly different states of the object.
- Discrepancies: The discrepancies in the system are eventually reconciled, ensuring eventual consistency.

Dynamo uses two operations in its simple interface for storing objects associated with a key: *get()* and *put()*:

1. The *get(key)* operation finds the object replicas, which are associated with the key in the storage system, returning either a single object or alternatively a list of objects with a context.
2. The *put(key, context, object)* operation is used to determine where the replicas of the object are to be placed on the basis of the associated key writing the replicas to storage.

System metadata about the object is encoded. It is opaque to the caller and includes, for example, the version of the object. The context information is stored with the object in such a way that it is possible for the system to verify the validity of the context object in the put request. Both the key and the object are delivered by the caller in the form of an opaque byte array. The system applies a MD5 hash on the key and generates a 128-bit identifier. This is used to determine the storage nodes responsible for the serving of the key.

Dynamo is based on self-organization and emergence: nodes are identical, they can emerge and disappear, and the data will be automatically balanced in the ring. Unlike popular commercial data stores, Dynamo exposes issues of data consistency and reconciliation logic to the developers.

Dynamo also uses a full membership scheme, with each node aware of the data hosted by other nodes. Therefore, each node actively keeps the other nodes in the system informed of the full routing table using gossip protocol. In smallish environments this is enough, but scaling the design to run with tens of thousands of nodes is nontrivial because the overhead needed to maintain the routing table is a function of the system size.

Many different applications at Amazon use Dynamo with differing configurations. The exact way of configuring Dynamo for an application depends on the tolerance to delays or data discrepancy. The current main uses of Dynamo are:

- Business logic–specific reconciliation, where each data object will be replicated in multiple nodes (for example, the shopping cart service).
- Timestamp-based reconciliation, which is like the previous one but with a different reconciliation mechanism (for example, the service that maintains customers' session information).
- High-performance read engine, where Dynamo data store is built to be always writeable but some services are using it as a read engine (for example, services that maintain a product catalogue and promotional items).

Porting applications to use Dynamo should be relatively simple if the applications are designed to handle different modes of failure and other inconsistencies that may arise. Analysis is needed during the initial stages of the development to choose the appropriate conflict resolution mechanisms.

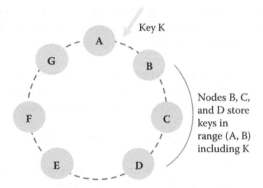

FIGURE 10.2
Overview of the ring structure in Dynamo.

10.1.2 Ring Membership

The membership changes of nodes in the Dynamo storage ring are initiated by administrators, both to join a node or remove a node. Figure 10.2 illustrates the ring structure. The explicit mechanism was chosen to prevent transient outages causing unnecessary rebalancing, repair, or startup operations. The membership change and its timestamp are written to a persistent store, and a gossip-based protocol is used to propagate this information. This leads eventually to a consistent list of membership. With the gossip mechanism, every node in the ring regularly contacts a random ring node, reconciling the membership changes. These contacts are also used to reconcile the token mappings (mappings of virtual nodes in the hash space), which are stored persistently. Thus, information of partitioning and placement propagates via the gossip-based protocol as well. Every node will be aware of the token ranges of its peers and is therefore able to forward read/write operations to the proper set of nodes.

If an administrator joins two nodes to the ring successively, the ring could be temporarily in logically partitioned state before the new nodes hear of each other. Therefore Dynamo uses seed nodes, obtained from static configuration or a configuring service. All nodes will eventually reconcile with seeds, effectively preventing logical partitions.

10.1.3 Partitioning Algorithm

Incremental scalability is one of Dynamo's foremost design requirements. A mechanism is needed to partition the data dynamically over the set of nodes (which are the storage hosts) of the system. The partitioning scheme in Dynamo uses consistent hashing for distributing the load across multiple hosts. The output range of a hash function is basically a fixed circular ring, and the largest hash value is wrapped around to the smallest. As discussed in Chapter 4, in a consistent hashing scheme:

- The nodes of the system are assigned random values representing positions on the ring.
- A data item identifiable by a key is attached to a node by first hashing the key of the data item in order to yield its position X on the ring and then stepping the ring clockwise to find the first node with a position larger than X.

In this way, each node occupies the local responsibility for a region in the ring lying between the node and its predecessor on the ring. Thus, in consistent hashing the departure or arrival of a node only affects its adjacent neighbors.

However, Dynamo must consider some challenges present in consistent hashing algorithms, like the random position assignment of nodes on the ring, leading to nonuniform data and load distribution and the obliviousness of the algorithm to the performance heterogeneity of nodes.

Dynamo uses its own variant of consistent hashing. Nodes are assigned to multiple points in the ring instead of the practice of mapping a node to a single point. To do this, Dynamo uses virtual nodes, and a new node added to the system is assigned to multiple positions in the ring.

There are three advantages in Dynamo's usage of virtual nodes:

- The potential unavailability of a node has the consequence that the failed node's load is evenly dispersed among the remaining nodes.

- A newly emergent node (reentering or fresh creation) gets an equivalent amount of load from the existing nodes.

- The system can allow for the heterogeneity of the infrastructure and make a capacity-based decision on the number of the other nodes for which a node is responsible.

10.1.4 Replication

Dynamo replicates its data on multiple hosts. This is done because of the need for both high availability and durability. Each data item is replicated at N hosts, where N is an instance parameter. Each key, k, is assigned to a coordinator node, handling the replication of the data items within its range. The coordinator locally stores each key within its range but also replicates the keys in the ring at the $N-1$ clockwise successor nodes. Thus each node is responsible for the ring region between itself and its Nth predecessor.

The list of nodes responsible for a particular key is called its preference list. Every node in the system is able to determine the identity of the nodes that should be in the list for any specific key. Because there could be node failures, the preference list must contain more than N nodes. However, the use of virtual nodes makes it possible that the first N successive positions for a key could be owned by less than N separate nodes. Therefore, a key's preference list must contain only distinct physical nodes.

10.1.5 Data Versioning

Dynamo uses a mechanism by which updates are propagated to all replicas asynchronously, providing a measure of eventual consistency. Eventual consistency means that there could be immediate inconsistencies: if a *put()* call returns to the caller before all replicas have been updated. This could cause situations where a subsequent *get()* operation returns an object without the latest updates.

In the case of no failures, there will be an upper bound on the update propagation times. However, it is possible that under, for example, server outages all replicas are not updated for an extended time period. All categories of Amazon applications are not vulnerable to such inconsistencies, however. For example, the shopping cart application can operate under these conditions. It only requires that an add-to-cart operation will be never forgotten or rejected. If a user makes changes to an older version of the cart when the current state is unavailable, the change is still meaningful in the cart environment and must be retained. Still, the change should not supersede the current state of the cart, which, while unavailable, could contain preservable changes.

In the Dynamo environment, add to cart and delete item from cart operations are carried through with *put()* requests. Additions to and removals from the cart are executed in the available version of the cart, and the different versions are adjusted and mediated later. Therefore, all modifications are treated in Dynamo as new and immutable versions of the object. This idea makes it possible for multiple versions of an object to exist simultaneously in the storage system.

Generally, older versions are subjugated by the newer versions. The system is capable of deciding which one is the authoritative version by syntactic reconciliation. In case of failures combined with concurrent updates, some version branching could happen, leading to conflicting versions of an object. Then a semantic reconciliation is performed, because the system is unable to reconcile the differing versions of the target object. In practice, the client must collapse the conflicting branches of data back into one. This can happen, for example, with the shopping cart when different versions are merged. Dynamo guarantees that an add to cart operation is never lost.

However, this is not necessarily the case with delete operation, and erased items may then emerge back in the cart. Thus there really might be several different versions of a single shopping cart in existence. This danger must be perceived when designing applications on top of the Dynamo system platform.

10.1.6   Vector Clocks

Vector clocks (lists of node/counter pairs) are used to detect causality between multiple object versions [193]. All object versions are guaranteed to have an associated vector clock. The vector clocks are central in facilitating the checking of a multiplicity of objects. Examining them will resolve the parallelity or the causal order of the branches.

The logic of ancestral relations goes like this: when comparing the clocks of a pair of objects, if we find that the counters in the object A clock are less-than-or-equal-to all of the nodes in the object B clock, then A is an ancestor of B and can be erased. If this is not the result, A and B are in conflict requiring reconciliation. A client obtains the object context from a read operation. The vector clock information is included in the context. When updating an object, the client passes the context to the system; this effectively specifies the version of object that is updated.

When Dynamo gets a read request, it will notice the multiple branches. If these are in conflict, Dynamo returns all objects at the leaves, the context including the version information. If an update is received using this context, the system will deem that all versions are reconciled and consequently the differing branches are collapsed into one.

The Dynamo process using vector clocks can be clarified with a simple example:

1. A new object $T1$ is written by a client. The node A handling the write increases its sequence number and uses this number to create the vector clock for $T1$ so that we have now object $T1$ and the associated clock $(A, 1)$.

2. Next the client updates $T1$. We assume that node A handles the update as well. The updated object is $T2$ and the associated clock is $(A, 2)$. $T1$ is the ancestor of $T2$, and therefore $T2$ overwrites $T1$. However, because of the asynchronous updating, there could well be replicas of $T1$ at nodes that have not updated to $T2$.

3. Next the same client updates $T2$ with a different server B handling the request. The system now has data $T3$ and the clock $[(A, 2), (B, 1)]$.

4. Now a different client reads $T2$, attempting to update it and node C writing the data. The system will have object $T4$ (descendant of $T2$) with clock $[(A, 2), (C, 1)]$.

After these updates, the situation is the following:

- If a node aware of $T1$ or $T2$ receives $T4$ with its context (clock), it is able to determine that $T1$ and $T2$ are obsolete and can be overwritten.
- If a node aware of $T3$ receives $T4$, it will find no causal relation between them.

Thus changes in $T3$ and $T4$ are not synchronized, and we must keep both versions of the data and ultimately present them to a client to be reconciled semantically.

If a client reads both $T3$ and $T4$, the context of the read will be a summary of the clocks of $T3$ and $T4$, resulting in $(A, 2)$, $(B, 1)$, $(C, 1)$. The client can now perform the reconciliation. If node A handles the write, A updates its sequence number in the version clock and the new data object $T5$ will have the clock $(A, 3)$, $(B, 1)$, $(C, 1)$.

Theoretically, the growing size of vector clocks might limit scalability, if a single object is handled by multiple servers. However, the write operations are in a normal case handled by the top N nodes in the preference list. Only when the case is a network partition or multiple server failure might handling be done by nodes outside the set of top N nodes. Then we might have to manage the problem of growing vector clocks. Dynamo is able to limit the size of vector clocks by using clock truncation where a timestamp is stored with each (node, counter) pair, indicating the last time the node updated the data item. If the number of (node, counter) pairs reaches a preset threshold, the oldest pair is erased from the clock.

10.1.7 Coping with Failures

The *get* and *put* operations can be used in Dynamo for all nodes and all keys. Both operations are invoked using Amazon's request processing HTTP framework, which is infrastructure-specific. A client can select a node using two strategies:

1. Routing the request through a generic load balancer. This will select a node on the basis of load information. In this case the client has no need to link Dynamo-specific code in application.
2. Using a client library that is partition-aware and routes requests directly to the co-ordinator nodes. The advantage in this case is the low latency achieved by skipping a possible forwarding step.

The coordinator node sitting within the top N nodes of the preference list handles read and write operations. Even if a request is routed to a random node through a load balancer, it will not become a coordinator node but will forward the request to a true, accessible coordinator among the top N nodes the preference list. This list is accessed from top down.

A consistency protocol is applied to maintain consistency in the replica set. In this protocol, two values can be configured for any successful read/write operation:

1. R is the minimum number of nodes for read.
2. W is the minimum number of nodes for write.

The requirement for R and W is $R + W > N$. This will produce a quorum-like system. Furthermore, both R and W are usually set to be less than N, because this generates better latency for operations such as put/get that are sensitive to the slowest replicas. Thus the following will happen:

- *put()* request: The receiving coordinator will generate the vector clock, do a local write, and send the new object with its context to the N accessible top nodes of the list. The write fails if fewer than $W - 1$ nodes respond.

- *get()* request: The coordinator will send a read for the given key to the topmost *N* reachable nodes in the preference list, waiting for *R* responses. If successful, it will return the result to the client. In case of multiple versions of the data, the coordinator evaluates the causality of the set and returns all causally unrelated versions. This leads to the reconciliation process, and eventually the older versions are overwritten by the reconciled version.

Dynamo has a failure-detection scheme to prevent attempts to communicate with unreachable nodes. Dynamo does not use a decentralized failure-detection protocol that would maintain a globally consistent view of failures. It is sufficient to use explicit local node join and leave methods because these inform nodes of permanent additions and removals. Furthermore, any temporary failed nodes are detected after unsuccessful communication attempts.

10.2   Overlay Video Delivery

Much of the expected IP traffic increase in the coming years will come from the delivery of video data in various forms [79]. Video delivery on the Internet will see a huge increase, and the volume of video delivery in 2013 is expected to be 700 times the capacity of the US Internet backbone in 2000. The study anticipates that video traffic will account for 91% of all consumer traffic in 2013. Therefore, solutions are needed on multiple protocol layers to be able to cope with this demand. The solutions need to be ISP friendly in the sense that unneccessary interdomain traffic is minimized [341]. In this chapter, we focus on overlay solutions for video delivery across the Internet.

10.2.1   Live Streaming

Although most peer-to-peer (P2P) solutions, such as Gnutella, BitTorrent, and Freenet, focus on file sharing, a number of P2P systems have been developed for live (real-time) media streaming [35]. These systems are also called P2PTV systems. These streams are typically TV channels [211]. In contrast to typical streaming systems, P2PTV users assist each other in downloading the stream. Each peer thus contributes upload capacity to the P2P network.

P2P has recently become a popular alternative to Content Delivery Network (CDN) technologies in order to address the growing demands for video traffic [248]. We can take PPLive as an example, which is a P2P-based video startup in China. PPLive[1] has demonstrated the use of 10 Mbps server distribution bandwidth to simultaneously serve 1.48 million users at a total consumption rate of 592 Gbps [326]. This has been realized by user-assisted streaming, in which the peers contribute resources to the P2P networks.

The aims of file-sharing and live streaming P2P differ. The former has the goal of high piece retrieval rate and does not consider the ordering of the pieces. For example, BitTorrent uses either rarest piece first or random piece-selection models. The latter has to balance the retrieval rate with the media playback rate. The file-sharing algorithms require that the file transfer completes before actions can be performed on the data. This is contrasted by the live streaming systems, which operate on the data at runtime.

Live streaming systems require more intricate piece-selection algorithms that support sequential retrieval of the pieces (in-order delivery). In order to balance efficiency with retrieval rate, small deviations are typically allowed from the sequential order. Figure 10.3

---

[1] www.pplive.com

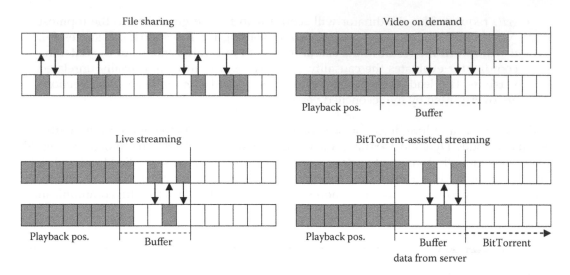

FIGURE 10.3
Comparison of piece-selection strategies.

illustrates the difference between the four types of media delivery: file sharing, VoD, live streaming, and BitTorrent-assisted streaming.

The SplitStream system presented in Chapter 8 is an example of a DHT-based application-layer multicast protocol that can be used to deliver live media stream. Another example is *data-driven overlay network (DONet)*, a system for live media streaming [360]. This system does not have an explicit overlay structure but instead adaptively forwards data based on current supply and demand. A new version of the system, called *Coolstreaming*, uses a hybrid push-pull scheme [199].

10.2.2   Video-on-Demand

P2P has been applied extensively to on-demand streaming of stored media, typically VoD [12, 250]. This form of data delivery has stricter delivery requirements than file sharing; however, it does not require real-time guarantees for piece retrieval. The stored media file needs to be retrieved at a rate that allows the pieces to be played back in a sequential order at the media playback rate.

The VoD delivery consists of two phases, namely the transfer phase and the playback phase. The transfer phase starts first, and the two phases are partially overlapping. A peer joins the system as a downloader and contacts other peers in order to download parts of a video. After a prebuffering period, the peer starts playback. When the video has finished playing, the peer may depart. If the peer is done downloading the video before playback is finished, it will stay as a seeder until it departs [233].

In P2P VoD systems, peers typically stay in the system as long as it takes to download the media file. During this time they contribute upload bandwidth to the system and share the pieces that they have. A peer may also choose to be altruistic and keep the file available after playback, thus becoming a seed in BitTorrent terminology for the file.

BitTorrent-assisted Streaming System (BASS)   BASS [104] is a hybrid system that uses a traditional client-server VoD streaming solution, which is extended with BitTorrent. BitTorrent is used to download non-time-sensitive data. The BitTorrent's rarest-first algorithm is modified to retrieve pieces that are after the current position in the playback. Thus retrieved pieces are stored for later playback. Pieces are downloaded in sequence from a dedicated media server, with the exception of pieces that have already been downloaded using BitTorrent or are being downloaded and are expected to be retrieved before the playback deadline.

BiToS   The BiToS system considers techniques for enhancing BitTorrent for view-as-you-download service. This is essentially VoD using a variant of BitTorrent. A peer can start to watch a video file while it is downloading. The motivation for VoD is clear for the peer: playback can start immediately rather than waiting for the whole file download to complete.

The BiToS system identifies the piece-selection mechanism as the only component that needs to be changed in order to support VoD. The key idea is to make the piece-selection algorithm aware of the playing order to enable smooth playback. However, this is not enough. The selection algorithm also has to support parallel downloading of pieces, the default being the rarest first order. The BiToS system strikes a balance between these two ways of ordering pieces.

The functional components of BiToS are the following:

- Received pieces, which contain all the downloaded pieces of the video stream. A piece state can have one of three values: downloaded, not downloaded or missed. A piece will be assigned the missed status if it did not meet its playback deadline.

- High-priority set, which contains the pieces of the video stream that have not been downloaded yet, are not missed, and are near being played back. This results in these pieces having higher priority over the other pieces.

- Remaining pieces set, which contains the pieces that have not been downloaded, are not missed, and are not in the high priority set.

With BiToS, a peer chooses with some probability p to download a piece of the video, which is in the high-priority set. The peer then has a probability of $1-p$ for a piece contained in the remaining pieces set. This probability p denotes the balance between the need to have pieces for immediate playback and future needs. An advantage of this approach is that the parameter p can be adjusted during runtime [334].

VoD Piece Selection Algorithms   The two well-known BitTorrent piece-selection algorithms suitable for VoD are BiToS [233] and the algorithm by Shah and Paris [289]. Both of these selection algorithms can be characterized as being window-based. They keep the basic piece-selection strategy of BitTorrent intact but restrict piece retrieval to a certain portion of the file. This window slides forward in the file as the download progresses. The two algorithms have two major differences: the definition of the window and whether the pieces are also requested from outside the window or not.

The starting point of the window is defined as the first piece that has not yet been downloaded and has not missed its playback deadline. Both BiToS and the Shah and Paris algorithm define a window size that is measured in the number of pieces. The difference in window definitions between the two is that, when calculating which pieces are within the window, the fixed-size window algorithm accepts both arrived and nonarrived pieces, whereas BiToS only accepts nonarrived pieces.

Give-to-Get   The *Give-to-Get* is a P2P VoD algorithm that discourages free-riding by letting peers favor uploading to other peers that have proven to be good uploaders. As a consequence, free-riders are only tolerated as long as there is spare capacity in the system. Simulation studies indicate that even if 20% of the peers are free-riders, Give-to-Get provides good performance to the well-behaving peers [233]. The Give-to-Get algorithm has been implemented in the Tribler P2P system.[2]

10.3   SIP and P2PSIP

The session initiation protocol (SIP) [276] is a text-based application-layer control protocol that can be used to establish, maintain, and terminate calls between two or more end points. The driving application for SIP has been telephony—e.g., the ability to be able to establish audio, such as *voice-over-IP (VoIP)* calls, or video sessions between mobile devices. SIP can be used to implement many of the advanced call-processing features found in the *Signaling System 7 (SS7)* used in traditional telecom systems. The SIP architecture has grown over the years and consists of a collection of IETF RFCs, Internet drafts, and best practices. In November 2000, SIP was accepted as a 3GPP signaling protocol, and therefore it is a key part of the *IP Multimedia Subsystem (IMS)* architecture for IP-based multimedia services.

Figure 10.4 presents an overview of the SIP architecture. The architecture consists of the user agents and proxy servers. SIP is a control-plane protocol and is used to set up sessions between user agents and also between user agents and servers. The main task of a SIP proxy is to mediate messages and to resolve a user's *address of record (AoR)* to the current IP address of a user. This is typically done by using DNS and a location server. SIP therefore offers a level of indirection that can be used to support various kinds of mobility, including session mobility [287]. The *session description protocol* is used to describe session properties. Typically *real-time transport protocol (RTP)* or some other protocol is used for transferring the actual content.

Decentralized VoIP calls can be seen as an emerging usage scenario. IETF has established a working group to develop protocols that can use the SIP protocol in networks where there are no centralized servers. The SIP architecture presented above requires the fixed-network proxies to function. Decentralization is desirable for mobile ad hoc networks that can be quickly deployed—for example, in case of emergency. The *P2P session initiation protocol working group (P2PSIP WG)*[3] is chartered to develop protocols and mechanisms for the use of the SIP protocol in environments where the service of establishing and managing sessions is handled by a collection of endpoints rather than centralized servers [24, 173, 215, 295].

The P2PSIP architecture is based on collections of nodes called P2PSIP peers and P2PSIP clients. P2PSIP peers define a distributed namespace in which overlay users are identified and provide mechanisms for locating users or resources within the P2PSIP overlay. P2PSIP clients differ from P2PSIP peers because they only utilize the overlay for discovery and do not contribute resources to the overlay. The overlay provides an alternative resolution service for the peers to the standardized SIP discovery process (RFC 3263).

Figure 10.5 illustrates the P2PSIP architecture. Instead of having dedicated fixed-network proxies, the discovery and message routing happens through a decentralized overlay—for example, a ring DHT. Each node in the DHT is responsible for part of the SIP proxy functionality. The processing load of the control plane signaling is therefore distributed over the P2P network.

---

[2] www.tribler.org
[3] tools.ietf.org/wg/p2psip/

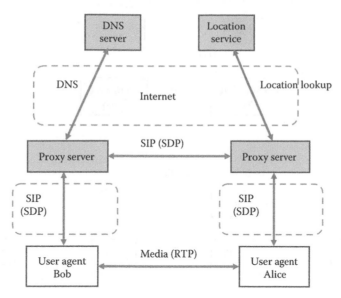

FIGURE 10.4
Overview of SIP architecture.

Figure 10.6 presents a comparison of the standard SIP and P2PSIP. The figure has two parts, the top and bottom diagrams. The top diagram presents the standard SIP call routing in which proxies are used to discover resources and route messages. The diagram shows the various resolution messages that are needed in order to find the destination—namely, DNS resolution and contacting the location server and database. The bottom diagram presents the P2PSIP call routing scenario, which contrasts with the standard SIP call routing process. In this case, the overlay (DHT) is used for both routing messages and for discovering the end points. Therefore DNS and the location server are not needed.

Several different P2PSIP protocols have been proposed, such as the RELOAD, SOSIM-PLE [223], decentralized SIP [221], DHT plug-ins [172], and P2PNS [24]. *Resource location and discovery (RELOAD)* is a P2P signaling protocol for use on the Internet. The protocol

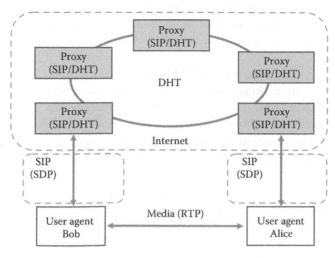

FIGURE 10.5
Overview of P2PSIP architecture.

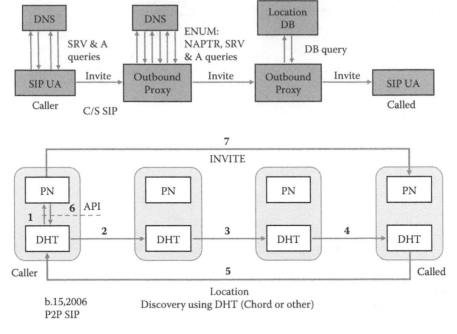

FIGURE 10.6
Comparison of SIP and P2PSIP.

provides its clients with an abstract storage and messaging service between a set of cooperating peers that form the overlay network. RELOAD is designed to support a P2PSIP network but can be utilized by other applications as well. The protocol has a security model based on identities obtained using a certificate enrollment service. The protocol also includes network address translation (NAT) traversal support.

The SIP usage of RELOAD allows SIP user agents to provide a P2P telephony service without requiring permanent proxies or registration servers. The RELOAD overlay itself performs the registration and rendezvous functions ordinarily associated with such servers. RELOAD is an overlay network that also offers storage capability. Records are stored under numeric addresses that are defined in the same identifier space as node identifiers. A node identifier determines the data items that the node is responsible for storing.

We briefly summarize the key features of RELOAD:

- Security framework: In the typical P2P network environment, peers do not necessarily trust each other. Therefore, a security framework is needed for building trust between peers. RELOAD uses a central enrollment server for granting credentials to peers. The credentials can then be used to authenticate and authorize peers.

- Usage model: The protocol has been designed to support a variety of signaling applications. These applications include P2P multimedia communications using SIP.

- NAT traversal: The protocol has been designed with the assumption that many network nodes will be behind NATs or firewalls. Thus NAT traversal support has been built into the RELOAD protocol. ICE is used to establish new protocol connections.

- High-performance routing: The distributed processing is distributed among peers in a P2P network. This means that the protocol needs to be lightweight.

- Pluggable overlay algorithms: The protocol has an abstract interface to the overlay layer, which can be used to support a variety of different structure and unstructured overlay algorithms. The specification defines how RELOAD is used with the Chord DHT, which is a mandatory part of the protocol; however, other algorithms can also be supported.

The security model of RELOAD is based on each node having one or more public key certificates. The protocol supports both certificates obtained from a central enrollment server and self-generated and self-signed certificates. The P2PSIP node identifier is computed as a digest of the public key. When self-certified identifiers are used, the system is vulnerable to a number of attacks, such as the Sybil and Eclipse attacks. Through the use of certificates, security is provided on three layers, namely the connection, message, and object levels. In the first level, connections between peers are secured using transport-layer security (TLS or DTLS). In the second level, messages are signed. In the third level, stored objects must be signed by the storing peer.

RELOAD distinguishes between clients and peers. A client is a an end system that uses the same protocol as the peers but is not required to participate in the DHT. Peers, on the other hand, are responsible for contributing resources to the overlay and running it. A client uses either the peer that is responsible for the client's identifier or an arbitrary peer in the overlay. The latter option is provided because in many environments it is not possible for a client to directly communicate with the designated peer due to issues pertaining to firewalls and NATs. The peers are required to support three overlay maintenance operations, namely *join*, *update*, and *leave*. The implementation of these operations is left to the DHT algorithm being used.

10.4   CDN Solutions

CDNs have evolved to improve Web site scalability and reliability. The first-generation systems mostly supported static or dynamic Web pages. With the second-generation CDNs, the focus has shifted to media delivery, such as streaming and VoD. CDNs have an important role in supporting the network in content delivery. The Internet was designed based on the end-to-end principle, which places the intelligence at the edges rather than the core of the network. This means that the core is optimized for packet forwarding. CDNs extend the end-to-end data-transport capability of the network by introducing techniques for the optimization of content delivery. Typical techniques used by CDN systems include server load balancing, caching, request routing, and content services.

10.4.1   Overview

The main components of a CDN architecture are content providers, CDN provider, and end users [43]. A content provider delegates the uniform resource identifier (URI) namespace to be replicated and distributed and buys this service from the CDN provider, or uses a collaborative open CDN such as the Coral CDN. The CDN provider has surrogate servers in various geographical locations and can thus distribute the requested resources efficiently. Clients are then directed to a surrogate server based on various metrics. In practice, a CDN is a collection of geographically distributed data centers.

The two general approaches for building CDNs are the overlay and network approach. In the former and frequently used approach, application-specific servers and caches distributed over the network manage content replication and delivery. In this approach, the

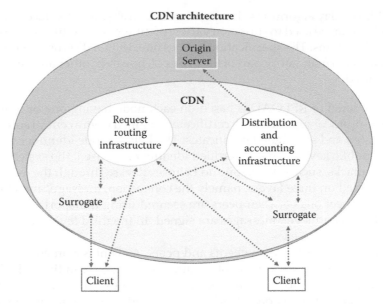

FIGURE 10.7
Content distribution networks.

basic network elements, namely routers, are not aware of the content delivery. In the latter, the network provides special support for content delivery. This approach is applicable within smaller networks; however, the requirement for custom network elements makes it very difficult to deploy in the wide-area environment.

Figure 10.7 presents as an example the central components of CDN systems. The example CDN architecture consists of an origin server that is the original source of data. The idea is that the CDN helps the origin server to distribute the data in an efficient and low-cost manner. In order to realize this distribution, the CDN has two important parts, namely the *request routing infrastructure* and the *distribution and accounting infrastructure*. The former is responsible for handling client-issued data queries. The clients need to be forwarded to a suitable surrogate or cache. The latter infrastructure is responsible for distributing the data given by the source across the Internet using surrogates and other caches. The surrogates store the data, and a record of this is made in the request routing infrastructure so that queries can be forwarded properly.

In addition to distribution, accounting is also needed to keep track of how data is accessed across the CDN. In Chapter 2 we discussed the costs of internetworking and observed that *internet service providers (ISPs)* have a motivation to minimize excessive interdomain traffic, especially through tier-1 transit. CDNs offer a convenient way to reduce this interdomain traffic by distributing the data based on anticipated and current demand.

Taxonomy   Figure 10.8 illustrates the key aspects of CDNs. A hosting CDN simply makes data available and does not offer relaying services. Therefore a hosting CDN includes the origin server. A relaying CDN offers either full or partial site content delivery service for an external origin server.

A request-routing system routes requests from clients to surrogate servers. CDNs use a variety of proprietary techniques to direct clients to the surrogate servers, including DNS-based request routing, URL rewriting, application or network level anycast, and HTTP-based redirection.

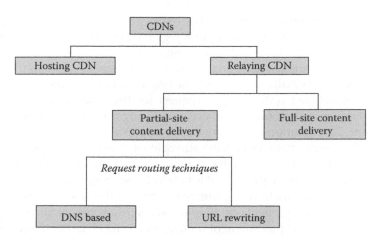

FIGURE 10.8
CDN taxonomy.

The system needs to be able to take into account various metrics when deciding to which surrogate server a request is forwarded—for example, network proximity, distance, client perceived latency, and replica server and data center load. The delivery technique used by a CDN has implications for the request-routing system. The two frequently used delivery techniques are full-site and partial-site content delivery. In the former, the CDN replicates a whole site. In the latter, only certain resources in a site are replicated and handled by the CDN infrastructure. In this case, the resource-specific redirection can be realized using either DNS-based request routing or URL rewriting. Figure 10.9 illustrates DNS-based request direction in which a DNS resolution first consults the content provider's DNS, which refers the DNS resolver to the CDN's DNS. The CDN can then direct the request to the proper surrogate server.

The full-site content delivery model is simple; however, it requires additional solutions for dynamic content. Both models require a content outsourcing solution, which is typically cooperative push-based, noncooperative pull-based, or cooperative pull-based. In the first

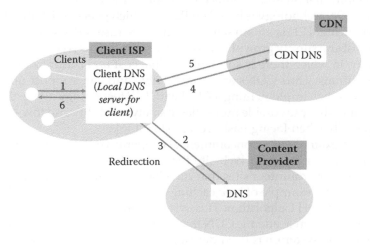

FIGURE 10.9
DNS-based request direction.

content outsourcing solution, content is pushed to the surrogate servers by the origin server. Given that surrogate servers cooperate, a greedy-global heuristic replication algorithm can be used to optimize content placement. The second solution is used by most popular CDN services. This approach directs clients to the closest surrogate servers. In the case of cache miss, the surrogate servers pull content from the origin server. This means that the creation of a new replica is delayed until the first request for the content is received. The third solution differs from noncooperative pull because surrogate servers cooperate to fetch the requested content. This solution is used by the Coral CDN.

The optimal placement of content on the surrogate servers is an important problem for CDNs. The key questions pertain to what content to replicate and where to replicate [72, 178, 305]. The problem of replica placement is to select K surrogate servers out of N possible sites such that an objective function is optimized. Typically the objective function takes into account the network topology, client population and content access patterns. The problem of determining the number and placement of replicas can be modeled as the center placement problem. Two related problems are the K-median problem and the facility location problem, which have been shown to be NP-hard [298].

The cached and replicated resources at surrogate servers have an expiration time after which they are considered to be stale. Different cache update techniques are employed by CDNs to ensure that the content is fresh. The most common of these techniques is the periodic update.

CDN Types   CDNs can be divided into three categories:

- Commercial
- Cooperative (federated)
- P2P-based overlays

Commercial CDNs offer content and service distribution. Examples of commercial CDNs include Akamai and LimeLight. Both use similar *domain name system (DNS)* redirection techniques to connect end clients with content servers. DNS redirection is used to reroute client requests to local clusters of machines, in many cases data centers. This rerouting behavior is influenced by detailed maps of the Internet topology, which are based on *border gateway protocol* information and various measurement methods.

An emerging CDN technology is based on P2P, in which peers assist the CDN infrastructure in load balancing. P2P is an attractive solution because it does not involve increased infrastructure cost and supply grows linearly with demand. This is contrasted by traditional CDN technology, in which there is substantial initial cost and centralized scheduling and replication algorithms. Thus CDNs are more reliable and can support quality-of-service parameters. Rather than implementing a CDN with P2P technology alone, the combination of the two appears to have favorable properties, namely reliability and low-cost incremental scalability, especially when facing flash crowds [135, 136].

CoralCDN is an example of a noncommercial cooperative P2P content distribution network that allows a user to run a Web site that offers high performance and scalability. The system is based on volunteer sites that run CoralCDN to automatically replicate content [137, 342]. Another example of Web caching using a DHT is the Squirrel system, which is based on the Pastry DHT algorithm. This system is intended for organization-wide networks [169]. CoDeeN is an academic CDN that provides caching of Web content and HTTP redirection [343]. The system has been developed on top of the global PlanetLab testbed and consists of a set of proxy servers that act both as request redirectors and surrogate servers. Globule is an open-source collaborative CDN that provides replication of content, server monitoring, and client request redirection to replicas. The internode latency is used as a proximity measure when forwarding requests and optimally placing replicas to clients. The system is implemented as an Apache HTTP server module [254].

Companies that operate CDNs invest in large-scale infrastructure, such as data centers, in order to be able to meet the demands for content distribution. The introduction of a new CDN service usually involves high investments, which motivates the development of CDN prototyping tools. One tool is the CDNSim, which is a CDN simulator designed to be a tool for predicting the behavior of CDN services in a controlled environment [304].

Performance Performance of a CDN pertains to the average and peak volume of traffic that can be sustained by the system. From the viewpoint of clients, latency also plays a crucial role. In general, five key metrics are used by the content providers to evaluate the performance of a CDN [43]:

- Cache hit ratio, which is defined as the ratio of the number of cached documents versus total documents requested
- Reserved bandwidth, which is the measure of the bandwidth used by the origin server
- Latency, which refers to the user-perceived response time
- Surrogate server utilization, which refers to the fraction of time during which the surrogate servers remain busy
- Reliability, which involves packet-loss measurements that are used to determine the reliability of a CDN

Charging CDN providers charge their customers based on the traffic volume. Thus a logging and accounting mechanism is a crucial part of a CDN architecture. Information pertaining to request-routing and content delivery needs to be collected and then processed for billing and charging. Key factors in influencing the price of a CDN service include the following:

- Number of surrogate servers
- Size of content replicated over surrogate servers
- Bandwidth cost
- Variation of traffic distribution

10.4.2 Akamai

Akamai is the market leader in CDN services and owns tens of thousands of servers across the world in order to serve content even in the flash crowd scenarios, in which a specific page or resource receives massive amounts of queries. As a solution to increasing content demand, the CDN infrastructure must be able to take geographical location of both requests and servers into account. Indeed, Akamai's approach is based on this observation [164].

The Akamai CDN uses DNS extensively to be able to connect end users to nearby surrogate servers. This is realized by hosting the content in some specific host name, for example under the Akamai domain. When an end user requests content that is available through the CDN, the URL of the resource will initially point at the service provider. The name is then resolved by the client's DNS resolver, and a new host name is obtained pointing to the CDN domain. The client then performs a secondary resolution to find an IP address for the surrogate server hosting the content. This secondary DNS query is processed by

Akamai's own private DNS infrastructure, which can then direct the query to the nearest surrogate. The resolution may return several IP addresses in order to allow client-side load balancing.

The Akamai load-balancing system uses border gateway protocol (BGP) information to determine network topology. This topology information is then combined with real-time network statistics to derive a dynamic view of the CDN infrastructure. The state of the surrogate servers is constantly monitored. Akamai also uses software agents that simulate end-user behavior to measure system latency and failure rates. The CDN system can then be provisioned based on these measurements—for example, the internal DNS system can distribute load by varying the surrogate server IP addresses that are returned to clients. Similarly, a top-level domain name server (NS) resolver can be instructed to direct traffic away from overloaded data centers.

10.4.3   Limelight

Limelight Networks entered the CDN market in 2001 with a vision to deliver a media-quality Internet experience to Internet users. The Limelight CDN provides distributed on-demand and live delivery of various kinds of media, including video. The system is built around data centers that host surrogate servers across the world, and it uses a similar DNS redirection technique to that used by Akamai. Web addresses are first mapped to one of Limelight's data centers and then to one of the surrogates. Unlike Akamai, Limelight allows customers direct use of CDN-based hostnames in their Web sites.

10.4.4   Coral

Coral CDN is a P2P CDN that allows users to run highly popular Web sites for the price of a regular broadband Internet connection. The idea is to use volunteer sites that run Coral CDN software and replicate content when users access it. The system has been designed to help Web sites cope with flash crowds. The system is easy to use, and it requires a small change in the hostname of an object's URL. A P2P DNS layer is used to redirect browsers to nearby cache nodes. This redirection effectively reduces the load on the origin Web server. The system uses a latency-optimized hierarchical indexing abstraction called *distributed sloppy hash table (DSHT)* [137, 138]. A content publisher can use CoralDNS by appending *.nyud.net:8090* to the hostname in a URL. DNS redirection is then used to direct clients to nearby Coral Web caches. The caches cooperate to minimize the load and latency of content access.

CoralCDN uses a key/value indexing infrastructure built on top of a DHT. The index allows nodes to locate nearby cached copies of content. It also mitigates flash crowds by distributing the requests across caches. The system design of the Coral DHT differs from the more traditional decentralized algorithms in that the architecture is based on clusters of nodes rather than nodes that are dispersed across wide areas. These clusters are exposed in the Coral interface to higher-level software, and they are part of the DNS redirection mechanism. In addition to this cluster-based nature, Coral also uses a weakened notion of consistency than what is typical for DHTs.

Each Coral node belongs to several distinct DSHT clusters. Each cluster has a diameter, which is the maximum preferred round-trip time (RTT) time. The system is parameterized by a fixed hierarchy of diameters that are called levels. Every node is a member of one DSHT at each level. A group of nodes can form a level-i cluster if a high-enough fraction of their pair-wise RTTs are below the level-i diameter threshold. The depth of a hierarchy can be arbitrary, but it can be expected to be a relatively small number (for example, 3). The higher-level and faster clusters are queries before slower lower-level clusters.

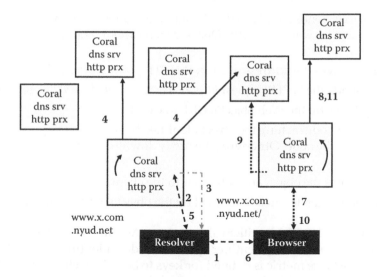

FIGURE 10.10
Overview of Coral.

Coral offers the following API for applications:

- *put(key, val, ttl, [levels])*: It inserts a mapping from the key to some value and spec-ifies a time-to-live for the mapping. The caller may optionally specify restriction of the operation to certain levels of the hierarchy.
- *get(key, [levels])*: It retrieves a subset of the values stored under a key. The caller may optionally specify restriction of the operation to certain levels of the hierarchy.
- *nodes(level, count, [target], [services])*: It returns the specified number (count) of neighbors belonging to the node's cluster at the specified levels. The caller may optionally specify the target IP address, which is then taken into account when selecting the neighbors. In this case, the neighbors should be near the given tar-get. The services option can be used to find only neighbors that run the given service.
- *levels()*: It returns the number of levels in the Coral's hierarchy and their RTT thresholds.

Figure 10.10 shows the steps for accessing a Coral-based URL—for example, *http://www.x.-com.nyud.net:8090/*. The two main stages are DNS redirection and HTTP request handling. Both of these stages involve the Coral indexing infrastructure.

1. A client consults its local DNS resolver for *www.x.com.nyud.net*.
2. The local resolver uses some Coral DNS servers, possibly starting from the top-level DNS servers.
3. A Coral DNS server receives the query and probes the client to determine the RTT and last few network hops.
4. The probe results are then used to find any known nameservers or HTTP proxies near the client.

5. The DNS server sends a reply to the client that includes any servers found through Coral. If no servers were found, the DNS server returns a random set of nameservers and proxies.

6. The client receives the address of a Coral HTTP proxy for *www.x.com.nyud.net*.

7. The client sends the HTTP request *http://www.x.com.nyud.net:8090/* to the proxy.

8. The proxy looks up the Web object's URL in Coral.

9. If Coral finds a node caching the object identified by the URL, the proxy downloads the object from the node. Otherwise, the proxy downloads the object from the origin server.

10. The proxy stores the object and returns it to the client.

11. The proxy stores a reference (identified by the object's URL) to itself in Coral.

The Coral DHT uses 160-bit identifier values as keys. As with many DHTs, node identifiers are defined in the same 160-bit identifier space. A node's identifier is the SHA-1 hash of its IP address. A distance metric is defined for keys to be able to cluster keys. Every DSHT node maintains a routing table that is used to find the closest node for any given key. The routing tables are based on Kademlia, which was presented in Chapter 5 and uses the XOR geometry. The *put* operation stores a key/value pair at a node closest to the key. Similarly, the *get* operation uses the routing tables to find the closest node that is responsible for the key.

The Coral algorithm routes messages to their destination (key) by visiting nodes whose distances to the key are approximately halved at each hop. Coral uses sloppy storage that caches key/value pairs at nodes that are close to the key in order to reduce congestion in the routing system. Frequent cached references to the same key can result in congestion in the shortcut nodes. The sloppy nature of Coral can be seen as a distinguishing feature that contrasts the typical DHT algorithms, which place key/value pairs to the node responsible for the closest key.

Figure 10.11 illustrates the Coral DHT routing algorithm. Each Coral node has the same identifier in all clusters. The node is placed in the same place in each of its clusters. The higher-level clusters are sparser than lower-level clusters. A node can be identified in a cluster by its shortest unique identifier prefix. The prefixes are stored in trees based on the XOR metric. A key observation is that routing is first performed using a higher-level cluster and then can be switched to a lower-level cluster on the fly. In the figure, a requesting node R initiates a query on its highest-level (level-2) cluster. This is done in order to find nearest replicas. The routing finds a node storing the requested key (1) and the value is returned to R. In this case, it is not necessary to consider the lower-layer clusters. If this routing in the highest-level does not produce a cache hit, the request will reach the node C_2, signifying the failure to locate the object at this level. The requesting node R then searches the level-1 cluster. R continues the search from C_2 on level-1 (3) because the identifier space up to the prefix of C_2 has already been covered. Eventually the search can switch to the global cluster (4). Even in this case, the search is efficient because lookups avoid testing the same identifier subspaces multiple times.

Sloppiness in the DSHT mitigates hotspots when nodes search for new clusters and test random subsets of nodes for acceptable RTT thresholds. Without sloppiness in the structure, hotspots could distort RTT measurements and limit scalability. The system allows the merging of clusters into the same namespace and the splitting of clusters into disjoint subsets, while minimizing oscillatory behavior. Merging can be started as a side-effect of a lookup to a node that has changed clusters. The notion of a cluster center provides a stable point about which to separate nodes.

Although the Coral CDN offers scalability and flexibility, centralized CDNs appear to offer two benefits over this system. The network measurement and monitoring in centralized

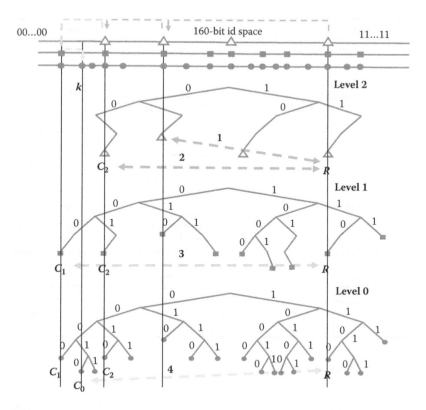

FIGURE 10.11
XOR-based routing in Coral.

CDNs can be seen to be more accurate and reflect the network topology better. Coral CDN does not rely on BGP topology information. Moreover, the system does not have information about node identities or their precise locations. The system offers less aggregate storage capacity than centralized CDNs, because the cache management is completely localized. Thus the approach appears to be more suitable for small organizations with static content. Indeed, performance measurements of Coral indicate that it allows under-provisioned Web sites to achieve significantly higher capacity.

10.4.5 Comparison

Figure 10.12 presents a comparison of the selected CDN technologies discussed in this chapter. Akamai, Limelight Networks, and many other commercial CDNs utilize a proprietary network of data centers distributed across the world in order to provide efficient content distribution. These systems can support both static and dynamic content, as well as various streaming and on-demand content types. The key architectural component is the request-routing system that is used to forward clients to nearest surrogate servers that cache the requested data. The problem of optimally placing surrogates and distributing the content over them is nontrivial and requires constant monitoring of the network, request patterns, and the load of the surrogates. Most CDNs use DNS-based redirection techniques because this can be implemented in a transparent fashion to clients and does not require changes to basic network infrastructure.

A number of academic experimental CDNs have been proposed, such as Coral, CoDeeN, and Globule. Many of the proposals are based on a DHT—for example, Coral. The aim is

CDN	Type	Coverage	Solutions
Akamai	Commercial CDN service including streaming data	Market leader	Edge platform for handling static and dynamic content, DNS-based request-routing
Limelight networks	Commercial On-demand distribution, live video, music, games, etc.	Surrogate servers in over 70 locations in the world	Edge-based solutions for content delivery, streaming support, custom CDN for custom delivery solutions, DNS-based request-routing
Coral	Academic Content replication based on popularity (on demand), addresses flash crowds	Experimental, hosted on PlanetLab	Uses a DHT algorithm (Kademlia), support for static content, DNS-based request-routing
CoDeeN	Academic testbed Caching of content and redirection of HTTP requests	Experimental, hosted on PlanetLab, collaborative CDN	Support for static content, HTTP direction Consistent hashing for mapping data to servers
Globule	Academic Replication of content, server monitoring, redirection to available replicas	Apache extension, Open-source collaborative CDN	Support for static content, monitoring services, DNS-based request-routing

FIGURE 10.12
Comparison of selected CDN technologies.

to be able to provide a decentralized and collaborative CDN in which clients contribute resources to the system. This would ideally allow efficient low-cost content delivery and a way to deal with flash crowds. PlanetLab is used to host many of these proposals. The P2P CDN solutions appear to be the most suitable for small sites because they do not provide guarantees on the collaborative CDN capacity. Moreover, they typically do not have as robust a monitoring infrastructure as commercial CDNs.

11

Conclusions

In this book we have examined a number of algorithms, structures, and systems for coping with vast amounts of information. We are currently in the exabyte era and entering the era of the zettabyte and beyond. Current IP networking trends include peer-to-peer (P2P), Internet broadcast, both Internet and commercial video-on-demand (VoD), and high-definition content. These trends contribute to the load on the network and require new solutions to keep the network cost efficient and manageable.

One key observation regarding the IP infrastructure is that it is very difficult to change. Another observation is that the end-to-end communication nature of the Internet, the very idea of placing intelligence at the edges, provides a natural building ground for overlay systems. Therefore, overlay technology aims to extend network features in a low-cost and deployable fashion. It is clear that if file sharing clients support each other, the service provider does not have to have massive infrastructure or bandwidth to provide the resources. Indeed, this is where the strength of P2P systems lie, this ability of realizing information delivery systems in a cooperative way, utilizing local resources in the system.

Given the scale of the Internet, currently peer-assisted service delivery is becoming popular in order to alleviate scalability and performance issues. The classic examples of P2P services are BitTorrent for bulk data delivery and Skype for VoIP telephone calls. Infrastructure-based services may also be peer-assisted, with the clients, peers, collaborating in order to make the service more efficient and scalable.

Overlays also have limitations, as we have observed in this book. For example, they introduce additional latency (stretch) into the communications, may violate organizational boundaries, may suffer from connectivity problems due to firewalls and NATs, and are susceptible to malicious nodes and other security problems. Many P2P systems are prone to the tragedy of the commons, in which most peers do not contribute to the system but only selfishly consume resources and services offered by the P2P network. The study of incentives regarding participation in P2P networks has attracted a lot of interest, and many solutions have been proposed. Trust management can be seen to be crucial for P2P networks. The EigenTrust reputation algorithm and the give-to-get algorithm are examples of solutions in this domain.

We examined structured, unstructured, and hybrid solutions. The routing structures come in various forms and shapes, and they can be deterministic or probabilistic. The geometry of an overlay imposes constraints on its structure. Structure is necessary in order to ensure scalability. Examples of early systems that were unstructured but later incorporated structure in order to be more scalable include Gnutella and Freenet. Freenet is also an interesting example of small-world routing, in which a small number of selectively picked shortcuts result in improved scalability and network diameter. Structure also plays a crucial role for DHTs, in which the geometry and the way routing tables are built and maintained determine the scalability.

Many P2P systems and overlay solutions use probabilistic filters, typically Bloom filters and variants, to be able to maintain a compact representation of data items. Bloom filters offer constant-time lookups and a trade-off between the size of the filter and false positives.

Since data items may need to be updated and removed, a number of Bloom filter variants have been proposed that support counting and deletion of elements. Bloom filters are used extensively by P2P software, such as Gnutella, and Web caching systems, such as Squid.

Although P2P technology can be seen as being very successful with large-scale deployments with BitTorrent, Gnutella, and Skype, to name some popular systems, it also has not resulted in a revolution in how distributed applications are developed and deployed. The client-server paradigm still prevails, and it is the dominant paradigm for Web applications. Indeed, there are only a few examples of advanced DHT algorithms being used in production systems. Of the older technologies, linear hashing and LH* are extensively used in cluster solutions.

The more recent pioneers of this area are Amazon and Google, which use sophisticated algorithms to be able to manage and distribute massive amounts of data. The Amazon Dynamo system is a key example in this book of a system in production that uses advanced solutions from distributed computing. On the other hand, Amazon Dynamo is still owned and operated by a single company. Other deployed examples of the more advanced DHT technologies include the Coral CDN and Kademlia.

Regarding the query expressiveness of P2P systems, we can say that unstructured systems are more expressive than structured systems, because they allow each peer to evaluate the queries and the network topology is not constrained by the query language. Content-based routing is a recent research area in which messages are forwarded based on their contents. This model is expressive but also has scalability limitations in terms of the load that routers can handle. Unstructured and structured networks can be combined to form hybrids, and, in a similar manner, content-based routing systems can use structured networks to improve efficiency and scalability.

One possible application of overlay technology for the Internet is in the form of a control plane. In traditional telecommunications systems, there is a clear separation between the control plane and the content plane. In the Internet, there is no such separation and all interactions are over IP, and UDP or TCP. Overlays provide a scalable way to implement the separation of the content and control planes over the Internet. The i3 overlay can be seen as an example system that could offer this kind of service. The challenge with this kind of a service would be in taking organizational boundaries into account in the overlay. Content-based systems can also be seen as candidates for a control-plane that supports expressive information routing. An unsolved problem with content-based routing systems is spam and how to prevent it.

Cloud computing aims to provide various functions and services over the Internet by exposing remotely invocable APIs or virtualized resources. The cloud-based applications and services are facilitated by the cloud infrastructure, which needs to be resilient and scalable. The term cloud computing encompasses various kinds of services and service infrastructures. Typically, the infrastructure involves communications, storage, data distribution, and security facilities. These facilities are often hosted by data centers and high-performance computing clusters.

Current cloud infrastructure providers include Amazon, Google, Microsoft, and Yahoo. Figure 11.1 illustrates the cloud infrastructure that spans over the Internet. Amazon Web Services provide services on a utility computing basis, and the infrastructure is based on scalable structures such as those used in the Dynamo platform. Google App Engine uses dedicated infrastructure to execute applications and store their data. App Engine supports two programming languages, namely Python and Java, and offers a limited set of APIs—for example, nonrelational data storage using the BigTable structure.

Overlay technology can been seen in a facilitating role in cloud computing and Internet services in general. In this book we have presented the basic building blocks of this new infrastructure that have been developed over the past decade. Many of the solutions can

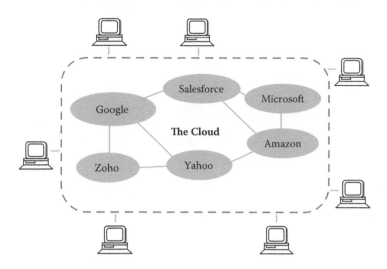

FIGURE 11.1
Cloud services.

also be applied for solutions on layers of the protocol other than the application layer. For example, Bloom filters can also be used on network layer routers. From the viewpoint of layering and network routers, overlay technology can be seen as a way to experiment with different kinds of alternatives for both intradomain and interdomain routing. Some of these solutions may then ultimately replace some of the current network layer technology in routers.

Wide-area overlays in the form of data-centric and content-based systems are not yet mainstream solutions, but they offer great promise for application developers and users alike in that routing, forwarding, and processing is performed in terms of the supply and demand for data. Indeed, the solutions covered in the book are paving the way toward information networking, in which the data supply and demand drives the network and offers good placement and distribution of data in terms of a number of parameters, including performance, accountability and organizational boundaries, security, and business models. There are many research questions still unanswered for information networking; however, the groundwork has already been done in terms of cluster-based systems and unstructured, structured, and hybrid decentralized overlays.

References

1. CIDR Report, 2009. http://www.cidr-report.org.
2. Shen Lin, Francois Taïani, and Gordon S. Blair. 2008. Facilitating gossip programming with the GossipKit framework. Lecture notes in computer science. In *DAIS* 5053:238–252. Oslo, Norway.
3. Masayuki Abe. 1998. Universally verifiable mix-net with verification work independent of the number of mix servers. In *Proceedings of EUROCRYPT 1998*, LNCS 1403, Espoo, Finland: Springer-Verlag.
4. Karl Aberer, Philippe Cudré-Mauroux, Anwitaman Datta, Zoran Despotovic, Manfred Hauswirth, Magdalena Punceva, and Roman Schmidt. 2003. P-Grid: A self-organizing structured P2P system. *SIGMOD Rec.* 32(3):29–33.
5. Lada A. Adamic. 2000. Zipf, power-law, Pareto—a ranking tutorial. *Technical Report*, Oct.: Information Dynamics Lab, HP Labs, Palo Alto, CA.
6. Lada A. Adamic and Bernardo A. Huberman. 2002. Zipf's law and the Internet. *Glottometrics* 3:143–150.
7. Marcos K. Aguilera, Robert E. Strom, Daniel C. Sturman, Mark Astley, and Tushar D. Chandra. 1999. Matching events in a content-based subscription system. In *PODC '99: Proceedings of the eighteenth annual ACM symposium on principles of distributed computing*, 53–61, New York: ACM Press.
8. Jalal Al-Muhtadi, Roy Campbell, Apu Kapadia, M. Dennis Mickunas, and Seung Yi. 2002. Routing through the mist: Privacy preserving communication in ubiquitous computing environments. In *ICDCS '02: Proceedings of the 22nd international conference on distributed computing systems (ICDCS'02)*, 74, Washington, DC: IEEE Computer Society.
9. David G. Andersen, Hari Balakrishnan, Nick Feamster, Teemu Koponen, Daekyeong Moon, and Scott Shenker. 2008. Accountable internet protocol (AIP). In *SIGCOMM '08: Proceedings of the ACM SIGCOMM 2008 conference on data communication*, 339–350, New York: ACM.
10. Ross J. Anderson, Roger M. Needham, and Adi Shamir. 1998. The steganographic file system. In *Proceedings of the 2nd international workshop on information hiding*, 73–82, London: Springer-Verlag.
11. Stephanos Androutsellis-Theotokis and Diomidis Spinellis. 2004. A survey of peer-to-peer content distribution technologies. *ACM Comp. Surv.* 36(4) (dec.):335–371.
12. Siddhartha Annapureddy, Saikat Guha, Christos Gkantsidis, Dinan Gunawardena, and Pablo Rodriguez. 2007. Exploring VoD in P2P swarming systems. In *INFOCOM*, 2571–2575. Anchorage, Alaska.
13. James Aspnes and Gauri Shah. 2007. Skip graphs. *ACM Trans. Algo.* 3(4):37.
14. Tuomas Aura. 2005. *Cryptographically generated addresses (CGA)*. RFC 3972, Mar.: IETF.
15. Baruch Awerbuch and Christian Scheideler. 2006. Towards a scalable and robust DHT. In *SPAA '06: Proceedings of the 18th annual ACM symposium on parallelism in algorithms and architectures*, 318–327, New York: ACM.
16. Robert Axelrod. 1984. *The evolution of cooperation*. New York: Basic Books.
17. Robert Axelrod. 1997. *The complexity of cooperation*: Princeton University Press. August.
18. Sebastien Baehni, Patrick Th. Eugster, and Rachid Guerraoui. 2004. Data-aware multicast. In *Proceedings of the 2004 international conference on dependable systems and networks (DSN 2004)*, 233–242. Florence, Italy.
19. Rena Bakhshi, Daniela Gavidia, Wan Fokkink, and Maarten Steen. 2009. An analytical model of information dissemination for a gossip-based protocol. In *ICDCN '09: Proceedings of the 10th international conference on distributed computing and networking*, 230–242, Berlin, Heidelberg: Springer-Verlag.

20. Hari Balakrishnan, Scott Shenker, and Michael Walfish. 2003. Semantic-free referencing in linked distributed systems. In *2nd international workshop on peer-to-peer systems (IPTPS '03)*, February. Berkeley, CA.

21. Roberto Baldoni, Carlo Marchetti, Antonino Virgillito, Roman Vitenberg. 2005. Content-based publishsubscribe over structured overlay networks. In *International conference on distributed computing systems (ICDCS 2005)*. 437–446. Columbus, OH.

22. Roberto Baldoni, Roberto Beraldi, Vivien Quema, Leonardo Querzoni, and Sara Tucci-Piergiovanni. 2007. TERA: Topic-based event routing for peer-to-peer architectures. In *DEBS '07: Proceedings of the 2007 inaugural international conference on distributed event-based systems*, 2–13, New York: ACM.

23. Lali Barrière, Pierre Fraigniaud, Evangelos Kranakis, and Danny Krizanc. 2001. Efficient routing in networks with long range contacts. In *DISC '01: Proceedings of the 15th international conference on distributed computing*, 270–284, London: Springer-Verlag.

24. Ingmar Baumgart. 2008. P2PNS: A secure distributed name service for P2PSIP. In *Proceedings of the 6th annual IEEE international conference on pervasive computing and communications (PerCom 2008)*, Mar., 480–485. Hong Kong, China.

25. Ingmar Baumgart, Bernhard Heep, and Stephan Krause. 2007. OverSim: A flexible overlay network simulation framework. In *Proceedings of the 10th IEEE global internet symposium (GI '07) in conjunction with IEEE INFOCOM 2007*, May, 79–84. Anchorage, AK.

26. J. Robert von Behren, Eric A. Brewer, Nikita Borisov, Michael Chen, Matt Welsh, Josh MacDonald, Jeremy Lau, and David E. Culler. 2002. Ninja: A framework for network services. In *ATEC '02: Proceedings of the general track of the annual conference on USENIX annual technical conference*, 87–102. Berkeley, CA: USENIX Association.

27. András Belokosztolszki, David M. Eyers, Peter R. Pietzuch, Jean Bacon, and Ken Moody. 2003. Role-based access control for publish/subscribe middleware architectures. In *Proceedings of the 2nd international workshop on distributed event-based systems (DEBS'03)*, ACM SIGMOD. San Diego, CA.

28. Ashwin R. Bharambe, Mukesh Agrawal, and Srinivasan Seshan. 2004. Mercury: Supporting scalable multi-attribute range queries. *SIGCOMM Comput. Commun. Rev.* 34(4):353–366.

29. Ruchir Bindal, Pei Cao, William Chan, Jan Medved, George Suwala, Tony Bates, and Amy Zhang. 2006. Improving traffic locality in BitTorrent via biased neighbor selection. In *ICDCS*, 66 (4–7 Jul.). Lisboa, Portugal.

30. Ken Birman. 2007. The promise, and limitations, of gossip protocols. *SIGOPS Oper. Syst. Rev.* 41(5):8–13.

31. Kenneth P. Birman, Mark Hayden, Oznur Ozkasap, Zhen Xiao, Mihai Budiu, and Yaron Minsky. 1999. Bimodal multicast. *ACM Trans. Comput. Syst.* 17(2):41–88.

32. Burton H. Bloom. 1970. Space/time trade-offs in hash coding with allowable errors. *Commun. ACM* 13(7):422–426.

33. Christian Böhm, Stefan Berchtold, and Daniel A. Keim. 2001. Searching in high-dimensional spaces: Index structures for improving the performance of multimedia databases. *ACM Comput. Surv.* 33(3):322–373.

34. Béla Bollobás and Oliver Riordan. 2004. The diameter of a scale-free random graph. *Combinatorica* 24(1):5–34.

35. Thomas Bonald, Laurent Massoulié, Fabien Mathieu, Diego Perino, Andrew Twigg. 2008. Epidemic live streaming: Optimal performance trade-offs. In *Proceedings of ACM SIGMETRICS* (2–6 Jun.), 325–336. Annapolis, Maryland.

36. Francois Bonnet, Anne-Marie Kermarrec, and Michel Raynal. 2007. Small-world networks: From theoretical bounds to practical systems. In *OPODIS*, vol. 4878 of *Lecture notes in computer science*, ed. Eduardo Tovar, Philippas Tsigas, and Hacne Fouchal, 372–385: Guadeloupe, French West Indies: Springer.

37. Flavio Bonomi, Michael Mitzenmacher, Rina Panigrahy, Sushil Singh, and George Varghese. 2006. Beyond bloom filters: From approximate membership checks to approximate state machines. In *SIGCOMM '06: Proceedings of the 2006 conference on applications, technologies, architectures, and protocols for computer communications*, 315–326, Pisa Italy: ACM.

38. Flavio Bonomi, Michael Mitzenmacher, Rina Panigrahy, Sushil Singh, and George Varghese. 2006. An improved construction for counting bloom filters. In *14th Annual European symposium on algorithms, LNCS 4168*, 684–695.

39. Giovanni Bricconi, Emma Tracanella, Elisabetta Di Nitto, and Alfonso Fuggetta. 2000. Analyzing the behavior of event dispatching systems through simulation. Lecture notes in computer science, In *HiPC* (Dec. 17–20), 131–140. Banglore, India.

40. Andrei Broder and Michael Mitzenmacher. 2002. Network applications of bloom filters: A survey. In *Internet Mathematics*, vol. 1, 636–646.

41. Andrei Z. Broder and Michael Mitzenmacher. 2001. Using multiple hash functions to improve IP lookups. In *INFOCOM (22–26 Apr.)*, 1454–1463. Anchorage, AL.

42. Ioana Burcea, Hans-Arno Jacobsen, Eyal de Lara, Vinod Muthusamy, and Milenko Petrovic. 2004. Disconnected operation in publish/subscribe middleware. In *Mobile data management*. Berkeley, CA.

43. Rajkumar Buyya, Mukaddim Pathan, and Athena Vakali, eds. 2008. *Content delivery networks (lecture notes electrical engineering)*: Springer-Verlag.

44. John Byers, Jeffrey Considine, Michael Mitzenmacher, and Stanislav Rost. 2002. Informed content delivery across adaptive overlay networks. In *SIGCOMM '02: Proceedings of the 2002 conference on applications, technologies, architectures, and protocols for computer communications*, 47–60, Pittsburgh, PA: ACM.

45. Luis Felipe Cabrera, Michael B. Jones, and Marvin Theimer. 2001. Herald: Achieving a global event notification service. In *Proceedings of the 8th workshop on hot topics in operating systems (HotOS-VIII)*. Elmau/Oberbayern, Germany.

46. Alexis Campailla, Sagar Chaki, Edmund Clarke, Somesh Jha, and Helmut Veith. 2001. Efficient filtering in publish-subscribe systems using binary decision diagrams. In *ICSE '01: Proceedings of the 23rd international conference on software engineering*, 443–452, Washington, DC: IEEE Computer Society.

47. Fengyun Cao and Jaswinder Pal Singh. 2005. MEDYM: Match-early with dynamic multicast for content-based publish-subscribe networks. In *Proceedings of the ACM/IFIP/USENIX 6th international middleware conference Nov. 28–Dec. 2 (Middleware 2005)*. Grenoble, France.

48. Fengyun Cao and Jaswinder Pal Singh. 2004. Efficient event routing in content-based publish-subscribe service networks. In *Proceedings of IEEE INFOCOM*, Mar. HongKong, China: IEEE.

49. Mauro Caporuscio, Paola Inverardi, Patrizio Pelliccione. 2002. Formal analysis of clients mobility in the Siena publish/subscribe middleware. *Technical Report*, Oct.: Department of Computer Science, University of Colorado.

50. Mauro Caporuscio, Antonio Carzaniga, and Alexander Wolf. 2002. An experience in evaluating publish/subscribe services in a wireless network. In *WOSP '02: Proceedings of the 3rd international workshop on software and performance*, 128–133. New York: ACM.

51. Mauro Caporuscio, Antonio Carzaniga, and Alexander L. Wolf. 2003. Design and evaluation of a support service for mobile, wireless publish/subscribe applications. *IEEE Trans. Software Eng.*, 29(12) (Dec.):1059–1071.

52. Damiano Carra, Giovanni Neglia, and Pietro Michiardi. 2008. On the impact of greedy strategies in BitTorrent networks: The case of BitTyrant. In *Peer-to-Peer Computing*, 311–320, 200. Aachen, Germany.

53. Antonio Carzaniga. 1998. *Architectures for an event notification service scalable to widearea networks*. PhD thesis, Politecnico di Milano, Milano, Italy.

54. Antonio Carzaniga, Jing Deng, and Alexander L. Wolf. 2001. *Fast forwarding for content-based networking. Technical Report*, Nov., CU-CS-922-01, Department of Computer Science, University of Colorado.

55. Antonio Carzaniga, David S. Rosenblum, and Alexander L. Wolf. 1999. *Interfaces and algorithms for a wide-area event notification service. Technical Report*, Oct. Revised May, CU-CS-888-99, Department of Computer Science, University of Colorado, 2000.

56. Antonio Carzaniga, David S. Rosenblum, and Alexander L. Wolf. 2000. *Content-based addressing and routing: A general model and its application. Technical Report*, Jan., CU-CS-902-00, Department of Computer Science, University of Colorado.

57. Antonio Carzaniga, David S. Rosenblum, and Alexander L. Wolf. 2001. Design and evaluation of a wide-area event notification service. *ACM Trans. Comp. Sys.* 19(3) (Aug.):332–383.

58. Antonio Carzaniga, Matthew J. Rutherford, and Alexander L. Wolf. 2004. A routing scheme for content-based networking. In *Proceedings of IEEE INFOCOM 2004*, Mar., Hong Kong, China: IEEE.

59. Antonio Carzaniga and Alexander L. Wolf. 2001, Content-based networking: A new communication infrastructure. In *Infrastructure for Mobile and Wireless Systems*, (Oct. 15), 59–68. Scottsdale, AZ.

60. Antonio Carzaniga and Alexander L. Wolf. 2003. Forwarding in a content-based network. In *Proceedings of ACM SIGCOMM*, Aug. 2003, 163–174, Karlsruhe, Germany.

61. Miguel Castro, Manuel Costa, and Antony Rowstron. 2004. Should we build Gnutella on a structured overlay? *SIGCOMM Comput. Commun. Rev.* 34(1):131–136.

62. Miguel Castro, Peter Druschel, Ayalvadi Ganesh, Antony Rowstron, and Dan S. Wallach. 2002. Secure routing for structured peer-to-peer overlay networks. *SIGOPS Oper. Syst. Rev.* 36(SI):299–314.

63. Miguel Castro, Peter Druschel, Y. Charlie Hu, and Antony Rowstron. 2002. Exploiting network proximity in distributed hash tables. In *Proceedings of the international workshop on future directions in distributed computing (FuDiCo)*, June. Bertinoro, Italy.

64. Miguel Castro, Peter Druschel, Anne-Marie Kermarrec, Animesh Nandi, Antony Rowstron, and Atul Singh. SplitStream: High-bandwidth multicast in a cooperative environment. 2003. In *19th ACM symposium on operating systems principles (SOSP'03)*, Oct. New York.

65. Miguel Castro, Peter Druschel, Anne-Marie Kermarrec, and Antony Rowstron. 2002. One ring to rule them all: Service discovery and binding in structured peer-to-peer overlay networks. In *EW10: Proceedings of the 10th workshop on ACM SIGOPS European workshop*, 140–145, New York: ACM.

66. Miguel Castro, Peter Druschel, Anne-Marie Kermarrec, and Antony Rowstron. 2002. Scribe: A large-scale and decentralized application-level multicast infrastructure. *IEEE J. Selected Areas in Commun. (JSAC)* 20(8) (Oct.).

67. Meeyoung Cha, Haewoon Kwak, Pablo Rodriguez, Yong-Yeol Ahn, and Sue B. Moon. 2007. I tube, you tube, everybody tubes: Analyzing the world's largest user generated content video system. In *Internet Measurement Conference*, (24–26 Oct.), 1–14. San Diego, CA.

68. Fay Chang, Jeffrey Dean, Sanjay Ghemawat, Wilson C. Hsieh, Deborah A. Wallach, Mike Burrows, Tushar Chandra, Andrew Fikes, and Robert E. Gruber. 2006. Bigtable: A distributed storage system for structured data. In *OSDI '06: Proceedings of the 7th USENIX symposium on operating systems design and implementation*, 5–15. Berkeley, CA: USENIX Association.

69. David L. Chaum. 1981. Untraceable electronic mail, return addresses, and digital 263 pseudonyms. *Commun. ACM* 24(2) (Feb.):84–90.

70. Yatin Chawathe, Sylvia Ratnasamy, Lee Breslau, Nick Lanham, and Scott Shenker. 2003. Making gnutella-like P2P systems scalable. In *SIGCOMM '03: Proceedings of the 2003 conference on applications, technologies, architectures, and protocols for computer communications*, 407–418, New York: ACM.

71. Bernard Chazelle, Joe Kilian, Ronitt Rubinfeld, and Ayellet Tal. 2004. The Bloomier filter: an efficient data structure for static support lookup tables. In *SODA '04: Proceedings of the 15th annual ACM-SIAM symposium on discrete algorithms*, 30–39, Philadelphia, PA: Society for Industrial and Applied Mathematics.

72. Yan Chen, Lili Qiu, Weiyu Chen, Luan Nguyen, Randy H. Katz, and Y. H. Katz. 2003. Efficient and adaptive web replication using content clustering. *IEEE J. Select. Areas Commun.* 21:979–994.

73. Yuan Chen, Karsten Schwan, and Dong Zhou. 2003. Opportunistic channels: Mobility-aware event delivery. In *Middleware (16–20 Jun.)*, 182–201. Rio de Janeiro, Brazil.

74. Gregory Chockler, Roie Melamed, Yoav Tock, and Roman Vitenberg. 2007. SpiderCast: A scalable interest-aware overlay for topic-based pub/sub communication. In *DEBS '07: Proceedings of the 2007 inaugural international conference on distributed event-based systems*, 14–25, New York: ACM.

75. David R. Choffnes and Fabián E. Bustamante. 2008. Taming the torrent: A practical approach to reducing cross-ISP traffic in peer-to-peer systems. *SIGCOMM Comput. Commun. Rev.* 38(4):363–374.

76. Nicolas Christin and John Chuang. 2005. A cost-based analysis of overlay routing geometries. In *INFOCOM 2005. 24th annual joint conference of the IEEE computer and communications societies. Proceedings IEEE* (13–17 Mar.) 4:2566–2577. Miami.

77. Nicolas Christin, Andreas S. Weigend, and John Chuang. 2005. Content availability, pollution and poisoning in file sharing peer-to-peer networks. In *EC '05: Proceedings of the 6th ACM conference on electronic commerce*, 68–77, New York: ACM.

78. Cisco. 2008. The exabyte era. http://www.hbtf.org/files/cisco_ExabyteEra.pdf, January.

79. Cisco. 2009. *Cisco visual networking index: Forecast and methodology, 2008–2013*, June: Cisco.

80. Dave Clark, Bill Lehr, Steve Bauer, Peyman Faratin, Rahul Sami, and John Wroclawski. 2006. Overlay networks and the future of the Internet. *Commun. Strategies* (63).

81. David D. Clark. 1988. The design philosophy of the DARPA internet protocols. In *SIGCOMM*, Aug., 106–114. Stanford, CA: ACM.

82. Ian Clarke, Scott G. Miller, Theodore W. Hong, Oskar Sandberg, and Brandon Wiley. 2002. Protecting free expression online with Freenet. *IEEE Internet Comput.* 6(1):40–49.

83. Ian Clarke, Oskar Sandberg, Brandon Wiley, and Theodore W. Hong. 2001. Freenet: A distributed anonymous information storage and retrieval system. In *International workshop on designing privacy enhancing technologies*, 46–66, New York: Springer-Verlag.

84. Edith Cohen and Scott Shenker. 2002. Replication strategies in unstructured peer-to-peer networks. In *SIGCOMM '02: Proceedings of the 2002 conference on applications, technologies, architectures, and protocols for computer communications*, 177–190, New York: ACM.

85. Reuven Cohen, Keren Erez, Daniel Ben-Avraham, and Shlomo Havlin. 2000. Resilience of the Internet to random breakdowns. *Phys. Rev. Lett.* 85(21) (Nov.):4626+.

86. Saar Cohen and Yossi Matias. 2003. Spectral bloom filters. In *SIGMOD '03: Proceedings of the 2003 ACM SIGMOD international conference on management of data*, 241–252, New York: ACM.

87. George Colouris, Jean Dollimore, and Tim Kindberg. 1994. *Distributed systems: Concepts and design*. 2nd ed. Boston, MA: Addison-Wesley.

88. Brian F. Cooper and Hector Garcia-Molina. 2005. Ad hoc, self-supervising peer-to-peer search networks. *ACM Trans. Inf. Syst.* 23(2):169–200.

89. Amy Beth Corman, Peter Schachte, and Vanessa Teague. 2007. QUIP: A protocol for securing content in peer-to-peer publish/subscribe overlay networks. In *ACSC '07: Proceedings of the 30th Australasian conference on computer science*, 35–40, Darlinghurst, Australia: Australian Computer Society, Inc.

90. Thomas H. Cormen, Charles E. Leiserson, and Ronald L. Rivest. 2001. *Introduction to algorithms*: The MIT Press.

91. Manuel Costa, Miguel Castro, Antony Rowstron, and Peter Key. 2004. PIC: Practical Internet coordinates for distance estimation. In *ICDCS '04: Proceedings of the 24th international conference on distributed computing systems (ICDCS'04)*, 178–187, Washington, DC: IEEE Computer Society.

92. Paolo Costa, Matteo Migliavacca, Gian Pietro Picco, Gianpaolo Cugola. 2003. Introducing reliability in content-based publish-subscribe through epidemic algorithms. In *Proceedings of the 2nd international workshop on distributed event-based systems (DEBS'03)*. San Diego, California.

93. Lenore J. Cowen. 1999. Compact routing with minimum stretch. In *SODA '99: Proceedings of the 10th annual ACM-SIAM symposium on discrete algorithms*, 255–260, Philadelphia, PA: Society for Industrial and Applied Mathematics.

94. Arturo Crespo, Orkut Buyukkokten, and Hector Garcia-Molina. 2003. Query merging: Improving query subscription processing in a multicast environment. *IEEE Trans. Knowl. Data Eng.* 15(1):174–191.

95. Arturo Crespo and Hector Garcia-Molina. 2002. Routing indices for peer-to-peer systems. In *ICDCS '02: Proceedings of the 22nd international conference on distributed computing systems (ICDCS'02)*, 23, Vienna, Austria: IEEE Computer Society.

96. Gianpaolo Cugola, Elisabetta Di Nitto, and Alfonso Fuggetta. 1998. Exploiting an event-based infrastructure to develop complex distributed systems. In *Proceedings of the 20th international conference on software engineering*, 261–270: IEEE Computer Society.

97. Gianpaolo Cugola and Hans-Arno Jacobsen. 2002. Using publish/subscribe middleware for mobile systems. *ACM SIGMOBILE Mobile Comput. Commun. Rev.* 6(4) (Oct.).

98. Gianpaolo Cugola and Gian Pietro Picco. 2005. *REDS: A reconfigurable dispatching system. In Technical Report, Politecnico di Milano*, Milan, Italy.

99. Gianpaolo Cugola, Elisabetta Di Nitto, and Gian Pietro Picco. 2000. Content-based dispatching in a mobile environment. In *Workshop su sistemi distribuiti: Algorithmi, architectture e linguaggi*, Sep. Ischia, Italy.

100. Gianpaolo Cugola, Davide Frey, Amy L. Murphy, and Gian Pietro Picco. 2004. Minimizing the reconfiguration overhead in content-based publish-subscribe. In *SAC '04: Proceedings of the 2004 ACM symposium on applied computing (Mar. 14–17)*, 1134–1140: Nicosia, Cyprus: ACM Press.

101. Frank Dabek, Russ Cox, Frans Kaashoek, and Robert Morris. 2004. Vivaldi: A decentralized network coordinate system. *SIGCOMM Comput. Commun. Rev.* 34(4):15–26.

102. Frank Dabek, Ben Zhao, Peter Druschel, John Kubiatowicz, and Ion Stoica. 2003. Towards a common API for structured peer-to-peer overlays. In *Proceedings of the 2nd international workshop on peer-to-peer systems (IPTPS03)*, Berkeley, CA.

103. Ayodele Damola, Victor Souza, Per Karlsson, and Howard Green. 2008. Peer-to-peer traffic in operator networks. In *Peer-to-peer computing*, 177–179.

104. Chris Dana, Danjue Li, David Harrison, and Chen-Nee Chuah. 2005. BASS: BitTorrent assisted streaming system for video-on-demand. In *IEEE 7th workshop on multimedia signal processing, (Oct. 30–Nov. 2)*, 1–4. Shanghai, China.

105. Giuseppe DeCandia, Deniz Hastorun, Madan Jampani, Gunavardhan Kakulapati, Avinash Lakshman, Alex Pilchin, Swaminathan Sivasubramanian, Peter Vosshall, and Werner Vogels. 2007. Dynamo: Amazon's highly available key-value store. *SIGOPS Oper. Syst. Rev.* 41(6):205–220.

106. Stephen E. Deering and Robert M. Hinden. 1998. Internet Protocol, Version 6 (IPv6) Specification.

107. Stephen E. Deering. 1989. Host extensions for IP multicasting.

108. Alan Demers, Dan Greene, Carl Hauser, Wes Irish, and John Larson. 1987. Epidemic algorithms for replicated database maintenance. In *Proceedings of the 6th annual ACM symposium on principles of distributed computing, (Aug. 10–12)*, 1–12. Vancouver, British Columbia, Canada.

109. Sarang Dharmapurikar, Praveen Krishnamurthy, Todd S. Sproull, and John W. Lockwood. 2004. Deep packet inspection using parallel bloom filters. *IEEE Micro* 24(1):52–61.

110. Sarang Dharmapurikar, Praveen Krishnamurthy, and David E. Taylor. 2003. Longest prefix matching using bloom filters. In *SIGCOMM '03: Proceedings of the 2003 conference on applications, technologies, architectures, and protocols for computer communications, (Aug. 25–29)*, 201–212, Karlsruhe, Germany: ACM.

111. Tim Dierks and Eric Rescorla. 2006. *The Transport Layer Security (TLS) Protocol Version 1.1*. RFC 4346, Apr.: IETF.

112. Roger Dingledine, Nick Mathewson, and Paul Syverson. 2004. Tor: The second-generation onion router. In *13th USENIX security symposium*, San Diego, CA.

113. Roger Dingledine, Michael J. Freedman, and David Molnar. 2001. Accountability. In *Peer-to-peer: Harnessing the benefits of a disruptive technology*: O'Reilly and Associates.

114. Roger Dingledine, Michael J. Freedman, and David Molnar. 2000. The free haven project: Distributed anonymous storage service. In *International workshop on designing privacy enhancing technologies*, LNCS. 67–95. Berkeley: Springer-Verlag.

115. John R. Douceur. 2002. The Sybil attack. In *IPTPS '01: Revised papers from the 1st international workshop on peer-to-peer systems*, 251–260, London: Springer-Verlag.

116. Peter Druschel and Antony I. T. Rowstron. 2001. Past: A large-scale, persistent peer-to-peer storage utility. In *HotOS VIII*, 75–80. Elmau/Oberbayern, Germany.

117. Peter Druschel and Antony Rowstron. 2001. Storage management and caching in PAST, a large scale, persistent peer-to-peer storage utility. In *18th ACM SOSP*. Lake Louise, Alberta, Canada.

118. Sérgio Duarte, José Legatheaux Martins, Henrique J. Domingos, and Nuno Preguia. 2001. DEEDS—A distributed and extensible event dissemination service. In *Proceedings of the 4th European research seminar on advances in distributed systems (ERSADS)*, Forli, Italy.

119. Jeremy Elson and Jon Howell. 2008. Handling flash crowds from your garage. In *ATC'08: USENIX 2008 annual technical conference*, 171–184, Berkeley, CA: USENIX Association.

120. Kave Eshghi. 2002. Intrinsic references in distributed systems. Technical Report, In *IEEE workshop on resource sharing in massively distributed systems*, 675–680.

121. Christian Esteve, Fábio L. Verdi, and Maurício F. Magalh aes. 2008. Towards a new generation of information-oriented internetworking architectures. In *CoNEXT '08: Proceedings of the 2008 ACM CoNEXT Conference*, 1–6, Madrid, Spain: ACM.

122. Patrick Th. Eugster, Pascal A. Felber, Rachid Guerraoui, and Anne-Marie Kermarrec. 2003. The many faces of publish/subscribe. *ACM Comput. Surv.* 35(2):114–131.

123. Patrick Th. Eugster and Rachid Guerraoui. 2002. Probabilistic multicast. 0:313–324. *International conference on dependable systems and networks* (23–26 Jun.) Bethesda, MD: IEEE Computer Society.

124. Patrick Th. Eugster, Rachid Guerraoui, S. B. Handurukande, Petr Kouznetsov, and Anne-Marie Kermarrec. 2001. Lightweight probabilistic broadcast. In *DSN '01: Proceedings of the 2001 international conference on dependable systems and networks (formerly: FTCS)*, 443–452. Washington, DC: IEEE Computer Society.

125. Francoise Fabret, H. Arno Jacobsen, Francois Llirbat, Joao Pereira, Kenneth Ross, and Dennis Shasha. 2001. Filtering algorithms and implementation for very fast publish/subscribe. In *Proceedings of the 20th international conference on management of data (SIGMOD 2001)*, ed. Timos Sellis and Sharad Mehrotra, 115–126. Santa Barbara, CA.

126. Michalis Faloutsos, Petros Faloutsos, and Christos Faloutsos. 1999. On power-law relationships of the Internet topology. In *SIGCOMM '99: Proceedings of the conference on applications, technologies, architectures, and protocols for computer communication*, 251–262, New York: ACM.

127. Li Fan, Pei Cao, Jussara Almeida, and Andrei Z. Broder. 1998. Summary cache: A scalable wide-area web cache sharing protocol. *SIGCOMM Comput. Commun. Rev.* 28(4):254–265.

128. Li Fan, Pei Cao, Jussara Almeida, and Andrei Z. Broder. 2000. Summary cache: a scalable wide-area web cache sharing protocol. *IEEE/ACM Trans. Netw.* 8(3):281–293.

129. Wu-chang Feng, Kang G. Shin, Dilip D. Kandlur, and Debanjan Saha. 2002. The BLUE active queue management algorithms. *IEEE/ACM Trans. Netw.* 10(4):513–528.

130. William Fenner. 1997. Internet group management protocol, version 2. RFC 2236 (Nov.).

131. Ludger Fiege, Felix C. Gärtner, Sidath B. Handurukande, and Andreas Zeidler. 2003. Dealing with uncertainty in mobile publish/subscribe middleware. In *1st international workshop on middleware for pervasive and ad-hoc computing (MPAC 03)*, Rio de Janeiro, Brazil.

132. Ludger Fiege, Felix C. Gärtner, Oliver Kasten, and Andreas Zeidler. 2003. Supporting mobility in content-based publish/subscribe middleware. Lecture notes in Computer Science, 2672. In *Middleware (Jun. 16–20)*, 103–122. Rio de Janerio, Brazil.

133. Geoffrey Fox and Shrideep Pallickara. 2002.The Narada event brokering system: Overview and extensions. In *Proceedings of the 2002 international conference on parallel and distributed processing techniques and applications (PDPTA'02)*, ed. H.R. Arabnia, 353–359: CSREA Press.

134. Pierre Fraigniaud and Philippe Gauron. 2006. D2B: A de Bruijn based content-addressable network. *Theor. Comput. Sci.*, 355(1):65–79.

135. Michael J. Freedman. 2007. Democratizing content distribution. PhD thesis, NYU.

136. Michael J. Freedman, Christina Aperjis, and Ramesh Johari. 2008. Prices are right: Managing resources and incentives in peer-assisted content distribution. In *Proceedings of the 7th international workshop on peer-to-peer systems (IPTPS08)*, Feb., Tampa Bay, FL.

137. Michael J. Freedman, Eric Freudenthal, and David Mazières. 2004. Democratizing content publication with coral. In *NSDI'04: Proceedings of the 1st conference on symposium on networked systems design and implementation*, 239–252. Berkeley, CA: USENIX Association.

138. Michael J. Freedman and David Mazières. 2003. Sloppy hashing and self-organizing clusters. In *Proceedings of the 2nd international workshop on peer-to-peer systems (IPTPS03)* Feb. Berkeley, CA.

139. Michael J. Freedman and Robert Morris. 2002. Tarzan: A peer-to-peer anonymizing network layer. In *Proceedings of the 9th ACM conference on computer and communications security (CCS 2002)*, Nov., Washington, DC.

140. Michael J. Freedman, Emil Sit, Josh Cates, and Robert Morris. 2002. Introducing Tarzan, a peer-to-peer anonymizing network layer. In *Proceedings of the 1st international 267 workshop on peer-to-peer systems (IPTPS02)*, Mar., Cambridge, MA.

141. Prasanna Ganesan, Krishna Gummadi, and Hector Garcia-Molina. 2004. Canon in G Major: Designing DHTs with hierarchical structure. In *ICDCS '04: Proceedings of the 24th international conference on distributed computing systems (ICDCS'04)*, 263–272. Washington, DC: IEEE Computer Society.

142. Prasanna Ganesan, Beverly Yang, and Hector Garcia-Molina. 2004. One torus to rule them all: Multi-dimensional queries in P2P systems. In *WebDB '04: Proceedings of the 7th international workshop on the web and databases*, 19–24. New York, NY: ACM.

143. Lixin Gao. 2001. On inferring autonomous system relationships in the internet. *IEEE/ACM Trans. Netw.* 9(6):733–745.

144. Cyril Gavoille and Stéphane Pérennès. 1996. Memory requirement for routing in distributed networks. In *PODC '96: Proceedings of the 15th annual ACM symposium on principles of distributed computing*, 125–133. New York: ACM.

145. Phillipa Gill, Martin Arlitt, Zongpeng Li, and Anirban Mahanti. 2007. Youtube traffic characterization: A view from the edge. In *IMC '07: Proceedings of the 7th ACM SIGCOMM conference on internet measurement*, 15–28. New York: ACM.

146. Christos Gkantsidis, Milena Mihail, and Amin Saberi. 2006. Random walks in peer-to-peer networks: Algorithms and evaluation. *Perform. Eval.* 63(3):241–263.

147. Ian Goldberg and David Wagner. 1998. TAZ servers and the rewebber network: Enabling anonymous publishing on the world wide web. *First Monday* 3(4) (Aug.).

148. David Goldschlag, Michael Reed, and Paul Syverson. 1999. Onion routing. *Commun. ACM* 42(2):39–41.

149. Steven D. Gribble, Eric A. Brewer, Joseph M. Hellerstein, and David Culler. 2000. Scalable, distributed data structures for internet service construction. In *OSDI'00: Proceedings of the 4th conference on symposium on operating system design & implementation*, 319–332. San diego, CA: USENIX Association.

150. Steven D. Gribble, Matt Welsh, Rob von Behren, Eric A. Brewer, David Culler, N. Borisov, S. Czerwinski, et al. 2001. The Ninja architecture for robust internet-scale systems and services. *Comput. Netw.* 35(4):473–497.

151. Björn Grönvall. 2002. Scalable multicast forwarding. *SIGCOMM Comput. Commun. Rev.* 32(1).

152. K. Gummadi, R. Gummadi, S. Gribble, S. Ratnasamy, S. Shenker, and I. Stoica. 2003. The impact of DHT routing geometry on resilience and proximity. In *SIGCOMM '03: Proceedings of the 2003 conference on applications, technologies, architectures, and protocols for computer communications*, 381–394. New York: ACM.

153. Krishna P. Gummadi, Richard J. Dunn, Stefan Saroiu, Steven D. Gribble, Henry M. Levy, and John Zahorjan. 2003. Measurement, modeling, and analysis of a peer-to-peer filesharing workload. In *SOSP '03: Proceedings of the 19th ACM symposium on operating systems principles*, 314–329. New York: ACM.

154. Lei Guo, Songqing Chen, Zhen Xiao, Enhua Tan, Xiaoning Ding, and Xiaodong Zhang. 2005. Measurement, analysis, and modeling of BitTorrent-like systems. In *Proceedings of the 5th ACM SIGCOMM conference on internet measurement*, 35–48. Berkeley, CA.

155. Abhishek Gupta, Ozgur Sahin, Divyakant Agrawal, and Amr El Abbadi. 2004. Meghdoot: Content-based publish:subscribe over P2P networks. In *Proceedings of the ACM/IFIP/USENIX 5th international middleware conference (Middleware '04)*. Toronto, Ontario, Canada.

156. Anjali Gupta, Barbara Liskov, and Rodrigo Rodrigues. 2003. One hop lookups for peer-to-peer overlays. In *HotOS'03: Proceedings of the 9th conference on hot topics in operating systems*, (May 18–21), 7–12. Lihue, Hawaii: USENIX Association.

157. Indranil Gupta, Ken Birman, Prakash Linga, Al Demers, and Robbert van Renesse. Kelips: 2003. Building an efficient and stable P2P DHT through increased memory and background overhead. In *Proceedings of the 2nd international workshop on peer-to-peer systems (IPTPS '03)*. (20–21 Feb.). Berkeley, CA.

158. Andrei Gurtov, Dmitry Korzun, Andrey Lukyanenko, and Pekka Nikander. 2008. Hi3: An efficient and secure networking architecture for mobile hosts. *Comput. Commun.* 31(10): 2457–2467.

159. Steven Hand and Timothy Roscoe. 2002. Mnemosyne: Peer-to-peer steganographic storage. In *IPTPS '01: Revised papers from the 1st international workshop on peer-to-peer systems*, 130–140. London: Springer-Verlag.

160. Nicholas J. A. Harvey, Michael B. Jones, Stefan Saroiu, Marvin Theimer, and Alec Wolman. 2003. SkipNet: A scalable overlay network with practical locality properties. In *USITS'03: Proceedings of the 4th conference on USENIX symposium on internet technologies and systems*, 113–126. Berkeley, CA: USENIX Association.

161. Steven Hazel, and Brandon Wiley. 2002. Achord: A variant of the chord lookup service for use in censorship resistant peer-to-peer publishing systems. In *1st international peer to peer systems workshop (IPTPS02)*. (Mar. 7–8). MIT Faculty Club, Cambridge, MA.

162. Joseph M. Hellerstein. 2003. Toward network data independence. *SIGMOD Rec.* 32(3):34–40.

163. Kevin Ho, Jie Wu, and John Sum. 2008. On the session lifetime distribution of Gnutella. *Int. J. Parallel Emerg. Distrib. Syst.* 23(1):1–15.

164. Cheng Huang, Angela Wang, Jin Li, and Keith W. Ross. 2008. Understanding hybrid CDNP2P: Why limelight needs its own red swoosh. In *NOSSDAV '08: Proceedings of the 18th international workshop on network and operating systems support for digital audio and video*, 75–80, New York: ACM.

165. Yongqiang Huang and Hector Garcia-Molina. 2001. Publish/subscribe in a mobile enviroment. In *Proceedings of the 2nd ACM international workshop on data engineering for wireless and mobile access*, 27–34:ACM.

166. Yongqiang Huang and Hector Garcia-Molina. 2004. Publish/subscribe in a mobile environment. *Wirel. Netw.* 10(6):643–652.

167. Daniel Hughes, Geoff Coulson, and James Walkerdine. 2005. Free riding on Gnutella revisited: The bell tolls? *IEEE Distributed Systems Online* 6(6):1.

168. IBM. 2002. *Gryphon: Publish/subscribe over public networks*, Dec. (White paper) http://researchweb.watson.ibm.com/distributedmessaging/gryphon.html.

169. Sitaram Iyer, Antony Rowstron, and Peter Druschel. 2002. Squirrel: A decentralized peer-to-peer web cache. In *Proceedings of the 21st symposium on principles of distributed computing (PODC)*, July, Monterey, CA.

170. John Jannotti, David K. Gifford, Kirk L. Johnson, M. Frans Kaashoek, and James W. O'Toole, Jr. 2000. Overcast: Reliable multicasting with on overlay network. In *OSDI'00: Proceedings of the 4th conference on symposium on operating system design & implementation*, 14–14, Berkeley, CA:USENIX Association.

171. Márk Jelasity, Alberto Montresor, and Ozalp Babaoglu. 2005. Gossip-based aggregation in large dynamic networks. *ACM Trans. Comput. Syst.* 23(3):219–252.

172. David A. Bryan, Bruce B. Lowekamp, and Cullen Jennings. 2005. SOSIMPLE: A serverless, standards-based, P2P SIP communication system. *First international workshop on advanced architectures and algorithms for internet delivery and applications, AAA-IDEA*, 15 June, 42–49, Orlando, FL.

173. Yang Ji, Chunhong Zhang, Lichun Li, Yao Wang, and Mao Tao. 2009. Architecture design of P2PSIP system. *Int. J. Distrib. Sen. Netw.* 5(1):85–85.

174. Xing Jin, Wanqing Tu, and S.-H. Gary Chan. 2009. Challenges and advances in using IP multicast for overlay data delivery. *IEEE Commun.* 47(6).

175. David B. Johnson, Charles E. Perkins, and Jari Arkko. 2004. *Mobility Support in IPv6.*, Jun., 269 [Standards Track RFC 3775]: IETF.

176. M. Frans Kaashoek and David R. Karger. 2003. Koorde: A simple degree-optimal distributed hash table. In *LNCS 2735*, 98–107. Berkeley, CA.

177. Sepandar D. Kamvar, Mario T. Schlosser, and Hector Garcia-Molina. 2003. The EigenTrust algorithm for reputation management in P2P networks. In *Proceedings of the 12th international world wide web conference*, 640–651. Budapest, Hungary: ACM.

178. Jussi Kangasharju, James W. Roberts, and Keith W. Ross. 2002. Object replication strategies in content distribution networks. *Comp. Commun.* 25(4):376–383.

179. David Karger, Eric Lehman, Tom Leighton, Rina Panigrahy, Matthew Levine, and Daniel Lewin. 1997. Consistent hashing and random trees: Distributed caching protocols for relieving hot spots on the world wide web. In *STOC '97: Proceedings of the 29th annual ACM symposium on theory of computing*, 654–663, New York: ACM.

180. David R. Karger and Matthias Ruhl. 2006. Simple efficient load-balancing algorithms for peer-to-peer systems. *Theor. Comp. Sys.* 39(6):787–804.

181. Jonas S. Karlsson, Witold Litwin, and Tore Risch. 1996. LH*LH: A scalable high performance data structure for switched multicomputers. In *EDBT '96: Proceedings of the 5th international conference on extending database technology*, 573–591, London:Springer-Verlag.

182. Sebastian Kaune, Tobias Lauinger, Aleksandra Kovacevic, and Konstantin Pussep. 2008. Embracing the peer next door: Proximity in Kademlia. In *Peer-to-Peer Computing*, 343–350. Aachen, Germany.

183. David Kempe, Jon Kleinberg, and Alan Demers. 2001. Spatial gossip and resource location protocols. In *STOC '01: Proceedings of the 33rd annual ACM symposium on theory of computing*, 163–172, New York: ACM.

184. Anne-Marie Kermarrec and Maarten van Steen. 2007. Gossiping in distributed systems. *SIGOPS Oper. Syst. Rev.* 41(5):2–7.

185. Dogan Kesdogan, Jan Egner and Roland Büschkes 1998. Stop-and-go-mixes providing probabilistic anonymity in an open system. In *Proceedings of information hiding workshop*, 83–98. Springer-Verlag.

186. Jon Kleinberg. 2000. The small-world phenomenon: An algorithm perspective. In *STOC '00: Proceedings of the 32nd annual ACM symposium on theory of computing (May 21–23)*, 163–170, Portland, OR: ACM.

187. Leonard Kleinrock and Farouk Kamoun. 1975. Hierarchical routing for large networks. *Comput. Net.* 1:155–174.

188. Miika Komu, Sasu Tarkoma, Jaakko Kangasharju, and Andrei Gurtov. 2005. Applying a cryptographic namespace to applications. In *Proceedings of the 1st ACM workshop on dynamic interconnection of networks (DIN 2005)*. Cologne, Germany.

189. Teemu Koponen, Mohit Chawla, Byung-Gon Chun, Andrey Ermolinskiy, Kye Hyun Kim, Scott Shenker, and Ion Stoica. 2007. A data-oriented (and beyond) network architecture. *SIGCOMM Comput. Commun. Rev.* 37(4):181–192.

190. Dmitri Krioukov, K. C. Claffy, Kevin Fall, and Arthur Brady. 2007. On compact routing for the internet. *SIGCOMM Comput. Commun. Rev.* 37(3):41–52.

191. John Kubiatowicz, David Bindel, Yan Chen, Steven Czerwinski, Patrick Eaton, Dennis Geels, Ramakrishna Gummadi, et al. 2000. OceanStore: An architecture for global-scale persistent storage. *SIGARCH Comput. Archit. News* 28(5):190–201.

192. Gu-In Kwon and John W. Byers. 2004. ROMA: Reliable overlay multicast with loosely coupled TCP connections. In *proceedings IEEE INFOCOM: The 23rd annual joint conference of the IEEE computer and communications societies (Mar. 7–11)*. 385–395. Hongkong, China.

193. Leslie Lamport. 1978. Time, clocks, and the ordering of events. *Commun. ACM, 21(7)* (Jul.):558–565.

194. Leslie Lamport, Robert Shostak, and Marshall Pease. 1982. The Byzantine generals problem. *ACM Trans. Program. Lang. Syst.* 4(3):382–401.

195. Arnaud Legout, Nikitas Liogkas, Eddie Kohler, and Lixia Zhang. 2007. Clustering and sharing incentives in BitTorrent systems. In *SIGMETRICS '07: Proceedings of the 2007 ACM SIGMETRICS international conference on measurement and modeling of computer systems*, 301–312, New York: ACM.

196. Shan Lei and Ananth Grama. 2004. Extended consistent hashing: An efficient framework for object location. In *ICDCS '04: Proceedings of the 24th international conference on distributed computing systems (ICDCS'04)*, 254–262,Washington, DC: IEEE Computer Society.

197. Kirill Levchenko, Geoffrey M. Voelker, Ramamohan Paturi, and Stefan Savage. 2008. Xl: An efficient network routing algorithm. In *SIGCOMM '08: Proceedings of the ACM SIGCOMM 2008 conference on data communication*, 15–26, New York: ACM.

198. Brian Neil Levine and Clay Shields. 2002. Hordes: A multicast based protocol for anonymity. *J. Comput. Secur.* 10(3):213–240.

199. Bo Li, Yang Qu, Gabriel Yik Keung, Susu Xie, Chuang Lin, Jiangchuan Liu, and Xinyan Zhang. 2008. Inside the new coolstreaming: Principles, measurements and performance implications. In *INFOCOM 2008. The 27th conference on computer communications, 1031–1039:* IEEE.

200. Guoli Li, Shuang Hou, and Hans-Arno Jacobsen. 2005. A unified approach to routing, covering and merging in publish/subscribe systems based on modified binary decision diagrams. In *ICDCS*, 447–457. Columbus, OH.

201. Ji Li, Karen Sollins, and Dah-Yoh Lim. 2005. Implementing aggregation and broadcast over distributed hash tables. *SIGCOMM Comput. Commun. Rev.* 35(1):81–92.

202. Jinyang Li, Jeremy Stribling, Robert Morris, M. Frans Kaashoek, and Thomer M. Gil. 2005. A performance vs. cost framework for evaluating DHT design tradeoffs under churn. In *INFOCOM*, 225–236. Miami, OH.

203. Zhi Li and Prasant Mohapatra. 2007. On investigating overlay service topologies. *Comput. Netw.* 51(1):54–68.

204. Witold Litwin, Marie-Anna Neimat, Gerard Levy, Yakham Ndiaye, and Mouhamed T. Seck. 1997. LH*s: A high-availability and high-security scalable distributed data structure. In *RIDE '97: Proceedings of the 7th international workshop on research issues in data engineering (RIDE '97) high performance database management for large-scale applications*, 141, Birmigham, England:IEEE Computer Society.

205. Witold Litwin. 1980. Linear hashing: A new tool for file and table addressing. In *VLDB '1980: Proceedings of the 6th international conference on very large data bases*, 212–223. VLDB Endowment. Montreal, Canada.

206. Witold Litwin, Jai Menon, and Tore Risch. 1998. *LH* schemes with scalable availability. Research Report* RJ10121 (91937): IBM.

207. Witold Litwin, Marie-Anne Neimat, and Donovan A. Schneider. 1994. RP*: A family of order preserving scalable distributed data structures. In *VLDB '94: Proceedings of the 20th international conference on very large data bases*, 342–353, San Francisco, CA: Morgan Kaufmann Publishers Inc.

208. Witold Litwin, Marie-Anne Neimat, and Donovan A. Schneider. 1996. LH*—A scalable, distributed data structure. *ACM Trans. Database Syst.* 21(4):480–525.

209. Witold Litwin, Hanafi Yakouben, and Thomas Schwarz. 2008. LH*RSP2P: A scalable distributed data structure for P2P environment. In *NOTERE '08: Proceedings of the 8th international conference on new technologies in distributed systems*, 1–6, New York: ACM.

210. Francesca Lo Piccolo, Giovanni Neglia, and Giuseppe Bianchi. 2004. The effect of heterogeneous link capacities in BitTorrent-like file sharing systems. In *Proceedings of HOT-P2P*, 40–47. Volendam, The Netherlands.

211. Richard John Lobb, Ana Paula Couto da Silva, Emilio Leonardi, Marco Mellia, and Michela Meo. 2009. Adaptive overlay topology for mesh-based P2P-TV systems. In *NOSS271 DAV '09: Proceedings of the 18th international workshop on network and operating systems support for digital audio and video*, 31–36, New York: ACM.

212. Boon Thau Loo, Ryan Huebsch, Ion Stoica, and Joseph M. Hellerstein. 2004. The case for a hybrid P2P search infrastructure. In *IPTPS (Feb. 26–27)*, 141–150. San Diego, CA.

213. Eng Keong Lua, Jon Crowcroft, Marcelo Pias, Ravi Sharma, and Steven Lim. 2005. A survey and comparison of peer-to-peer overlay network schemes. *IEEE Commun. Surv. Tut.* 7:72–93.

214. Qin Lv, Pei Cao, Edith Cohen, Kai Li, and Scott Shenker. 2002. Search and replication in unstructured peer-to-peer networks. In *SIGMETRICS '02: Proceedings of the 2002 ACM SIGMETRICS international conference on measurement and modeling of computer systems*, 258–259, New York:ACM.

215. Hongyan Ma and Baomin Xu. 2007. A hierarchical P2P architecture for SIP communication. In *NGMAST '07: Proceedings of the the 2007 international conference on next generation mobile applications, services and technologies*, 130–135, Washington, DC:IEEE Computer Society.

216. Lothar F. Mackert and Guy M. Lohman. 1986. R* optimizer validation and performance evaluation for distributed queries. In *VLDB*, 149–159. Kyoto, Japan.

217. Dahlia Malkhi, Moni Naor, and David Ratajczak. 2002. Viceroy: A scalable and dynamic emulation of the butterfly. In *PODC '02: Proceedings of the 21st annual symposium on principles of distributed computing*, 183–192, New York: ACM.

218. Balasubramaneyam Maniymaran, Marin Bertier, and Anne-Marie Kermarrec. 2007. Build one, get one free: Leveraging the coexistence of multiple P2P overlay networks. In *ICDCS '07: Proceedings of the 27th international conference on distributed computing systems*, 33, Toronto, Canada: IEEE Computer Society.

219. Gurmeet Singh Manku. 2003. Routing networks for distributed hash tables. In *PODC '03: Proceedings of the 22nd annual symposium on principles of distributed computing*, 133–142, New York: ACM.

220. Gurmeet Singh Manku, Mayank Bawa, and Prabhakar Raghavan. 2003. Symphony: Distributed hashing in a small world. In *USITS'03: Proceedings of the 4th conference on USENIX symposium on internet technologies and systems*, 10–10, Berkeley, CA: USENIX Association.

221. Jukka Manner, Simone Leggio, Tommi Mikkonen, Jussi Saarinen, Pekka Vuorela, and Antti Ylä-Jääski. 2008. Seamless service interworking of ad-hoc networks and the internet. *Comput. Commun.* 31(10):2293–2307.

222. Chip Martel and Van Nguyen. 2004. Analyzing Kleinberg's (and other) small-world models. In *PODC '04: Proceedings of the 23rd annual ACM symposium on principles of distributed computing*, 179–188, New York: ACM.

223. Isaias Martinez-Yelmo, Alex Bikfalvi, Ruben Cuevas, Carmen Guerrero, and Jaime Garcia. 2009. H-P2PSIP: Interconnection of P2PSIP domains for global multimedia services based on a hierarchical DHT overlay network. *Comput. Netw.* 53(4):556–568.

224. Laurent Massoulié and Milan Vojnovic. 2005. Coupon replication systems. In *Proceedings of ACM SIGMETRICS (Jun. 6–10)*, 2–13. Banff, Alberta, Canada.

225. Laurent Massoulié, Erwan Le Merrer, Anne-Marie Kermarrec, and Ayalvadi Ganesh. 2006. Peer counting and sampling in overlay networks: Random walk methods. In *PODC '06: Proceedings of the 25th annual ACM symposium on principles of distributed computing*, 123–132, New York: ACM.

226. Petar Maymounkov and David Mazières. 2002. Kademlia: A peer-to-peer information system based on the XOR metric. In *IPTPS*, 53–65. Cambridge, MA.

227. David Mazières and M. Frans Kaashoek. 1998. Escaping the evils of centralized control with self-certifying pathnames. In *EW 8: Proceedings of the 8th ACM SIGOPS European workshop on support for composing distributed applications*, 118–125, New York: ACM.

228. Alfred J. Menezes, Paul C. Van Oorschot, and Scott A. Vanstone. 1997. *Handbook of applied cryptography*. CRC Press.

229. Ralph C. Merkle. 1988. A digital signature based on a conventional encryption function. In *CRYPTO '87: A conference on the theory and applications of cryptographic techniques on advances in cryptology*, 369–378, London, UK: Springer-Verlag.

230. Michael Mitzenmacher. 2001. Compressed bloom filters. In *PODC '01: Proceedings of the 20th annual ACM symposium on principles of distributed computing*, 144–150, New York: ACM.

231. Paul V. Mockapetris and Kevin. J. Dunlap. 1988. Development of the domain name system. *SIGCOMM Comput. Commun. Rev.* 18(4):123–133.

232. Jacob Jan-David Mol, Johan A. Pouwelse, Dick H. J. Epema, and Henk J. Sips. 2008. Free-riding, fairness, and firewalls in P2P file-sharing. In K. Wehrle, W. Kellerer, S. K. Singhal, and R. Steinmetz, ed., *8th IEEE international conference on peer-to-peer computing*, Sep. 301–310: IEEE Computer Society.

233. Jacob Jan-David Mol, Johan A. Pouwelse, Michel Meulpolder, Dick H. J. Epema, and Henk J. Sips. 2008. Give-to-get: Free-riding-resilient video-on-demand in P2P systems. In *Multimedia Computing and Networking 818; SPIE 6818 (30–31 Jan.)*. San Jose, CA.

234. Gero Mühl. 2002. *Large-scale content-based publish/subscribe systems*. PhD thesis, Darmstadt University of Technology, Sep.

235. Gero Mühl. Disseminating information to mobile clients using publish/subscribe. *IEEE Internet Compu.* (May):46–53, 2004.

236. Gero Mühl and Ludger Fiege. 2001. Supporting covering and merging in content-based publish/subscribe systems: Beyond name/value pairs. *IEEE Distri. Sys. Online (DSOnline)* 2(7).

237. Gero Mühl, Ludger Fiege, Felix C. Gärtner, and Alejandro P. Buchmann. 2002. Evaluating advanced routing algorithms for content-based publish/subscribe systems. In A. Boukerche, S. K. Das, and S. Majumdar, eds., *The 10th IEEE/ACM international symposium on modeling, analysis and simulation of computer and telecommunication systems (MASCOTS 2002)*, (Oct.), 167–176, Fort Worth, TX: IEEE Press.

238. Jochen Mundinger, Richard R. Weber, and Gideon Weiss. 2008. Optimal scheduling of peer-to-peer file dissemination. *J. Scheduling* 11:105–120.

239. Jochen Mundinger, Richard Weber, and Gideon Weiss. 2006. Analysis of peer-to-peer file dissemination. *SIGMETRICS Perform. Eval. Rev.* 34(3):12–14.

240. Vinod Muthusamy and Milenko Petrovic. 2005. Publisher mobility in distributed publish/subscribe systems. In *4th international workshop on distributed event-based systems (DEBS'05)*, June, 421–427, Columbus, Ohio: IEEE Press.

241. Vinod Muthusamy, Milenko Petrovic, and Hans-Arno Jacobsen. 2005. Effects of routing computations in content-based routing networks with mobile data sources. In *MobiCom '05:*

Proceedings of the 11th annual international conference on mobile computing and networking, 103–116, New York: ACM.

242. Alper Tugay Mýzrak, Yuchung Cheng, Vineet Kumar, and Stefan Savage. 2003. Structured superpeers: Leveraging heterogeneity to provide constant-time lookup. In *WIAPP '03: Proceedings of the the 3rd IEEE workshop on internet applications*, 104, Washington, DC: IEEE Computer Society.

243. Pekka Nikander, Jukka Ylitalo, and Jorma Wall. 2003. Integrating security, mobility, and multi-homing in a HIP way. In *Proceedings of network and distributed systems security symposium (NDSS03)*, (Feb.), 87–89. San Diego, CA.

244. Object Management Group. 2001. *CORBA notification service specification v.1.0*, Mar.

245. Object Management Group. 2004. *Wireless access and terminal mobility in CORBA v.1.1*, Apr.

246. Anna Ostlin and Rasmus Pagh. 2003. Uniform hashing in constant time and linear space. In *STOC '03: Proceedings of the 35th annual ACM symposium on theory of computing*, 622–628, New York: ACM.

247. Jitendra Padhye, Victor Firoiu, Donald F. Towsley, and James F. Kurose. 2000. Modeling TCP Reno performance: A simple model and its empirical validation. *IEEE/ACM Trans. Netw.* 8(2):133–145.

248. Venkata N. Padmanabhan, Helen J. Wang, Philip A. Chou, and Kunwadee Sripanidkulchai. 2002. Distributing streaming media content using cooperative networking. In *NOSSDAV '02: Proceedings of the 12th international workshop on network and operating systems support for digital audio and video*, 177–186, New York: ACM.

249. Anna Pagh, Rasmus Pagh, and S. Srinivasa Rao. 2005. An optimal Bloom filter replacement. In *SODA '05: Proceedings of the 16th annual ACM-SIAM symposium on discrete algorithms*, 823–829, Philadelphia, PA: Society for Industrial and Applied Mathematics.

250. Nadim Parvez, Carey Williamson, Anirban Mahanti, and Nikas Carlsson. 2008. Analysis of BitTorrent-like protocols for on-demand stored media streaming. In *Proceedings of ACM SIGMETRICS (Jun. 2–6)*, 301–312. Annapolis, MD.

251. David Peleg and Eli Upfal. 1989. A trade-off between space and efficiency for routing tables. *J. ACM* 36(3):510–530.

252. Charles Perkins. 2002. *IP Mobility Support for IPv4*. IETF, Aug [Standards Track RFC 3344].

253. Gian Pietro Picco and Paolo Costa. 2005. Semi-probabilistic publish/subscribe. In *Proceedings of 25th IEEE international conference on distributed computing systems (ICDCS 2005)*. (575–585). Columbus, OH.

254. Guillaume Pierre and Maarten van Steen. 2006. Globule: A collaborative content delivery network. *IEEE Commun. Mag.* 44(8) (Aug.):127–133.

255. Peter Pietzuch and Jean Bacon. 2002. Hermes: A distributed event-based middleware architecture. In *Proceedings of the 1st international workshop on distributed event-based systems (DEBS'02)*. Vienna, Austria.

256. Peter R. Pietzuch. 2004. *Hermes: A scalable event-based middleware*. PhD thesis, Computer Laboratory, Queens' College, University of Cambridge, Feb.

257. C. Greg Plaxton, Rajmohan Rajaraman, and Andréa W. Richa. 1997. Accessing nearby copies of replicated objects in a distributed environment. In *SPAA '97: Proceedings of the 9th annual ACM symposium on parallel algorithms and architectures*, 311–320, New York: ACM.

258. Ivana Podnar and Ignac Lovrek. 2004. Supporting mobility with persistent notifications in publish/subscribe systems. In *3rd international workshop on distributed event-based systems (DEBS'04)*, May. Edinburgh, Scotland.

259. Ivana Podnar, Manfred Hauswirth, and Mehdi Jazayeri. 2002. Mobile push: Delivering content to mobile users. In *Proceedings of the 22nd international conference on distributed computing systems*, 563–570: IEEE Computer Society.

260. Alexandru Popescu, David Erman, Markus Fiedler, and Demetres Kouvatsosn. 2009. Routing in content addressable networks: Algorithms and performance. In *20th ITC specialist seminar (May 18–20)*: IEEE. Hoi An, Vietnam.

261. Jon Postel. *Simple mail transfer protocol*, RFC 2821, April 2001. http://www.rfc-editor.org/rfc/rfc2821.txt.

262. Dongyu Qiu and Weiqian Sang. 2008. Global stability of peer-to-peer file sharing systems. *Comp. Commun.* 31(2):212–219.

263. Donqyu Qiu and R. Srikant. Modeling and performance analysis of BitTorrent-like peer-to-peer networks. In *Proceedings of ACM SIGCOMM*, 367–378, 2004.

264. Changtao Qu, Wolfgang Nejdl, and Matthias Kriesell. 2004. Cayley DHTs—A group-theoretic framework for analyzing DHTs based on Cayley graphs. In *International symposium on parallel and distributed processing and applications (ISPA) LNCS 3358 (Dec. 13–15)*: Springer-Verlag. Hongkong, China.

265. Michael O. Rabin. 1989. Efficient dispersal of information for security, load balancing, and fault tolerance. *J. ACM* 36(2):335–348.

266. Venugopalan Ramasubramanian, and Emin Gün Sirer. 2004. Beehive: O(1)lookup performance for power-law query distributions in peer-to-peer overlays. In *NSDI'04: Proceedings of the 1st conference on symposium on networked systems design and implementation*, 8–8, Berkeley, CA: USENIX Association.

267. Sylvia Ratnasamy, Paul Francis, Mark Handley, Richard Karp, and Scott Schenker. 2001. A scalable content-addressable network. In *SIGCOMM '01: Proceedings of the 2001 conference on applications, technologies, architectures, and protocols for computer communications*, 161–172, San Diego, CA: ACM.

268. V. C. Ravikumar. 2005. EaseCAM: An energy and storage efficient TCAM-based router architecture for IP lookup. *IEEE Trans. Comput.* 54(5):521–533. Senior member– Rabi N. Mahapatra, and Fellow– Laxmi Narayan Bhuyan.

269. Michael K. Reiter and Aviel D. Rubin. 1997. *Crowds: Anonymity for web transactions. Technical Report.* . Available at: http://dimacs.rutgers.edu/TechnicalReports1997.html.

270. Patrick Reynolds and Amin Vahdat. 2003. Efficient peer-to-peer keyword searching. In *Middleware '03: Proceedings of the ACM/IFIP/USENIX 2003 international conference on middleware*, 21–40, New York: Springer-Verlag.

271. Sean Rhea, Dennis Geels, Timothy Roscoe, and John Kubiatowicz. 2004. Handling churn in a DHT. In *ATEC '04: Proceedings of the annual conference on USENIX annual technical conference*, 10–10, Berkeley, CA: USENIX Association.

272. Matei Ripeanu and Ian T. Foster. 2002. Mapping the Gnutella network: Macroscopic properties of large-scale peer-to-peer systems. In *IPTPS*, 85–93. Cambridge, MA.

273. Matei Ripeanu, Adriana Iamnitchi, and Ian Foster. 2002. Mapping the Gnutella network. *IEEE Internet Comput.* 6(1):50–57.

274. John Risson and Tim Moors. 2007. *Survey of research towards robust peer-to-peer networks: Search method,* Sep., RFC 4981: Internet Engineering Task Force.

275. Rodrigo Rodrigues, Barbara Liskov, and Liuba Shrira. 2002. The design of a robust peer-to-peer system. In *10th ACM SIGOPS European workshop*, Sep., Saint Emilion, France.

276. Jonathan Rosenberg, Henning Schulzrinne, Gonzalo Camarillo, Alan Johnston, Jon Peterson, Robert Sparks, Mark Handley, and Eve Schooler. 2002. SIP: Session initiation protocol, Jun., RFC 3261: IETF.

277. Antony Rowstron and Peter Druschel. 2001. Pastry: Scalable, decentralized object location and routing for large-scale peer-to-peer systems. In *IFIP/ACM international conference on distributed systems platforms (Middleware) (Nov. 12–16)*, 329–350. Heidelberg, Germany.

278. Antony Rowstron and Peter Druschel. 2001. Storage management and caching in PAST, a large-scale, persistent peer-to-peer storage utility. In *18th ACM symposium on operating systems principles (SOSP'01)*, Oct, 188–201. Chateau lake Louise, Banff, Canada.

279. Antony Rowstron and Peter Druschel. 2001. Storage management and caching in PAST, a large-scale, persistent peer-to-peer storage utility. *SIGOPS Oper. Syst. Rev.* 35(5):188–201.

280. Antony I., T. Rowstron and Peter Druschel. 2001. Pastry: Scalable, decentralized object location, and routing for large-scale peer-to-peer systems. In *Middleware 2001: 275 Proceedings of the IFIP/ACM international conference on distributed systems platforms Heidelberg*, 329–350, London: Springer-Verlag.

281. Antony I. T. Rowstron, Anne-Marie Kermarrec, Miguel Castro, and Peter Druschel. SCRIBE: The design of a large-scale event notification infrastructure. In *Networked Group Communication*, pages 30–43, 2001.

282. Jerome H. Saltzer, David P. Reed, and David D. Clark. 1984. End-to-end arguments in system design. *ACM TOCS* 2(4) (Nov.):277–288.

283. Oskar Sandberg. 2006. Distributed routing in small-world networks. In *Proceedings of the eigth workshop on algorithm engineering and experiments*. Miami, FL.

284. Stefan Saroiu, Krishna P. Gummadi, and Steven D. Gribble. 2002. A measurement study of peer-to-peer file sharing systems. In *Multimedia Computing and Networking (MMCN)*, Jan.

285. Stefan Saroiu, Krishna P. Gummadi, and Steven D. Gribble. 2003. Measuring and analyzing the characteristics of Napster and Gnutella hosts. *Multimedia Syst.* 9(2):170–184.

286. Cristina Schmidt and Manish Parashar. 2008. Squid: Enabling search in DHT-based systems. *J. Parallel Distrib. Comput.* 68(7):962–975.

287. Henning Schulzrinne and Elin Wedlund. 2000. Application-layer mobility using SIP. *SIGMOBILE Mob. Comput. Commun. Rev.* 4(3):47–57.

288. Andrei Serjantov. 2002. Anonymizing censorship resistant systems. In *IPTPS '01: Revised papers from the 1st international workshop on peer-to-peer systems*, 111–120, London: Springer-Verlag.

289. Purvi Shah and Jehan-Francois Paris. Peer-to-peer multimedia streaming using BitTorrent. 2007. In *Performance, computing, and communications conference, IPCCC 2007*, 340–347: IEEE International. New Orleans.

290. Adi Shamir. 1979. How to share a secret. *Commun. ACM* 22(11):612–613.

291. Kulesh Shanmugasundaram, Hervé Brönnimann, and Nasir Memon. 2004. Payload attribution via hierarchical bloom filters. In *CCS '04: Proceedings of the 11th ACM conference on computer and communications security*, 31–41, New York: ACM.

292. Kazuyuki Shudo, Yoshio Tanaka, and Satoshi Sekiguchi. 2008. Overlay weaver: An overlay construction toolkit. *Comput. Commun.* 31(2):402–412.

293. Atul Singh, Miguel Castro, Peter Druschel, and Antony Rowstron. 2004. Defending against eclipse attacks on overlay networks. In *EW11: Proceedings of the 11th workshop on ACM SIGOPS European workshop*, 21, New York: ACM.

294. Atul Singh, Tsuen-Wan Ngan, Peter Druschel, and Dan S. Wallach. 2006. Eclipse attacks on overlay networks: Threats and defenses. In *INFOCOM*. Barcelona. Spain.

295. Kundan Singh and Henning Schulzrinne. 2005. Peer-to-peer internet telephony using SIP. In *NOSSDAV '05: Proceedings of the international workshop on network and operating systems support for digital audio and video*, 63–68, New York: ACM.

296. Emil Sit and Robert Morris. 2002. Security considerations for peer-to-peer distributed hash tables. In *IPTPS '01: Revised papers from the first international workshop on peer-to-peer systems*, 261–269, London:Springer-Verlag.

297. Thirunavukkarasu Sivaharan, Gordon S. Blair, and Geoff Coulson. 2005. GREEN: A configurable and re-configurable publish-subscribe middleware for pervasive computing. In *Proceedings of DOA*. 732–749. Agia Napa, Cyprus.

298. Swaminathan Sivasubramanian, Michal Szymaniak, Guillaume Pierre, and Maarten van Steen. 2004. Replication for web hosting systems. *ACM Comput. Surv.* 36(3):291–334. http://www.globule.org/publi/RWHS_cs.html.

299. Nikolaos Skarmeas and Keith Clark. 1999. Content-based routing as the basis for intra-agent communication. In *Proceedings of the 5th international workshop on intelligent agents V: Agent theories, architectures, and languages (ATAL-98)* 1555 of *LNAI*, 345–362, Berlin: Springer.

300. Alex C. Snoeren, Craig Partridge, Luis A. Sanchez, Christine E. Jones, Fabrice Tchakountio, Beverly Schwartz, Stephen T. Kent, and W. Timothy Strayer. 2002. Single-packet IP traceback. *IEEE/ACM Trans. Netw.* 10(6):721–734.

301. Kunwadee Sripanidkulchai, Bruce Maggs, and Hui Zhang. 2004. An analysis of live streaming workloads on the internet. In *IMC '04: Proceedings of the 4th ACM SIGCOMM conference on Internet measurement*, 41–54, New York: ACM.

302. Pyda Srisuresh, Bryan Ford, and Dan Kegel. 2008. State of peer-to-peer (P2P) communication across network address translators (NATs). RFC 5128 (Informational), Mar.

303. Mudhakar Srivatsa and Ling Liu. Securing publish-subscribe overlay services with EventGuard. 2005. In *CCS '05: Proceedings of the 12th ACM conference on computer and communications security*, 289–298, New York: ACM.

304. Konstantinos Stamos, George Pallis, Dimitrios Katsaros, Athena Vakali, and Yannis Manolopoulos Antonis Sidiropoulos. 2009. CDNSim: A simulation tool for content distribution networks. *ACM transactions on modeling and computer simulation*, in press.

305. Konstantinos Stamos, George Pallis, Athena Vakali, and Marios D. Dikaiakos. 2009. Evaluating the utility of content delivery networks. In *UPGRADE-CN '09: Proceedings of the 4th edition of the UPGRADE-CN workshop on use of P2P, GRID and agents for the development of content networks*, 11–20, New York: ACM.

306. Angelos Stavrou, Dan Rubenstein, and Sambit Sahu. 2004. A lightweight, robust P2P system to handle flash crowds. *IEEE J. Select. Areas Commun.* 22(1):6–17.

307. Ion Stoica, Daniel Adkins, Shelley Zhuang, Scott Shenker, and Sonesh Surana. 2002. Internet indirection infrastructure. In *Proceedings of the 2002 conference on applications, technologies, architectures, and protocols for computer communications*, 73–86: ACM. Pittsburgh, PA.

308. Ion Stoica, Robert Morris, David Karger, Frans Kaashoek, and Hari Balakrishnan. 2001, Chord: A scalable peer-to-peer lookup service for internet applications. *Comp. Commun. Rev.* 31(4) (Oct.):149–160.

309. Ion Stoica, Robert Morris, David Liben-Nowell, David R. Karger, M. Frans Kaashoek, Frank Dabek, and Hari Balakrishnan. 2003. Chord: A scalable peer-to-peer lookup protocol for internet applications. *IEEE/ACM Trans. Netw.* 11(1):17–32.

310. Robert E. Strom, Guruduth Banavar, Tushar Deepak Chandra, Marc Kaplan, Kevan Miller, Bodhi Mukherjee, Daniel C. Sturman, and Michael Ward. 1998. Gryphon: An information flow based approach to message brokering. *CoRR*, cs.DC/9810019.

311. Daniel Stutzbach and Reza Rejaie. 2006. Improving lookup performance over a widely deployed DHT. In *INFOCOM*. Barcelona, Spain.

312. Daniel Stutzbach and Reza Rejaie. 2006. Understanding churn in peer-to-peer networks. In *IMC '06: Proceedings of the 6th ACM SIGCOMM conference on internet measurement*, 189–202, Pisa, Italy: ACM.

313. Daniel Stutzbach, Reza Rejaie, and Subhabrata Sen. 2008. Characterizing unstructured overlay topologies in modern P2P file-sharing systems. *IEEE/ACM Trans. Netw.* 16(2):267–280.

314. Sun Microsystems. 2001. *Java message service specification*, Jun.

315. Riikka Susitaival and Samuli Aalto. 2006. Modelling the population dynamics and the file availability in a BitTorrent-like P2P system with decreasing peer arrival rate. In *Proceedings of IWSOS*, 34–48, Passau, Germany.

316. Peter Sutton, Rhys Arkins, and Bill Segall. 2001. Supporting disconnectedness-transparent information delivery for mobile and invisible computing. In *CCGRID '01: Proceedings of the 1st international symposium on cluster computing and the grid*, 277, Brisbane, Australia: IEEE Computer Society.

317. Chunqiang Tang, Melissa J. Buco, Rong N. Chang, Sandhya Dwarkadas, Laura Z. Luan, Edward So, and Christopher Ward. 2005. Low traffic overlay networks with large routing tables. *SIGMETRICS Perform. Eval. Rev.* 33(1):14–25.

318. Chunqiang Tang, Zhichen Xu, and Sandhya Dwarkadas. 2003. Peer-to-peer information retrieval using self-organizing semantic overlay networks. In *SIGCOMM '03: Proceedings of the 2003 conference on applications, technologies, architectures, and protocols for computer communications*, 175–186, New York: ACM.

319. Chunqiang Tang, Zhichen Xu, and Mallik Mahalingam. 2003. pSearch: Information retrieval in structured overlays. *SIGCOMM Comput. Commun. Rev.*, 33(1):89–94.

320. Liying Tang and Mark Crovella. 2003. Virtual landmarks for the internet. In *IMC '03: Proceedings of the 3rd ACM SIGCOMM conference on internet measurement*, 143–152, New York: ACM.

321. Andreas Tanner and Gero Mühl. 2004. *A formalisation of message-complete publish/subscribe systems. Technical Report Rote Reihe*, Sep., Berlin University of Technology, Amsterdam, the Netherlands, October 2004. Brief announcement given at the 18th annual conference on distributed computing (DISC 2004).

322. Sasu Tarkoma. 2006. *Efficient content-based routing, mobility-aware topologies, and temporal subspace matching*. PhD thesis, Department of Computer Science, University of Helsinki. Available at ethesis.helsinki.fi.

323. Sasu Tarkoma. 2008. Dynamic filter merging and mergeability detection for publish/subscribe. *Pervasive and Mobile Comput.* 4(5):681–696.

324. Sasu Tarkoma and Jaakko Kangasharju. 2006. Optimizing content-based routers: Posets and forests. *Distri. Comput.* 19(1):62–77.

325. Sasu Tarkoma and Jaakko Kangasharju. 2007. On the cost and safety of handoffs in content based routing systems. *Comp. Net.* 51(6):1459–1482.

326. Ye Tian, Di Wu, Guangzhong Sun, and Kam-Wing Ng. 2008. Improving stability for peer-to-peer multicast overlays by active measurements. *J. Syst. Archit.* 54(1–2):305–323.

327. Peter Triantafillou and Andreas Economides. 2002. Subscription summaries for scalability and efficiency in publish/subscribe systems. In J. Bacon, L. Fiege, R. Guerraoui, A. Jacobsen, and G. Mühl, eds., *In Proceedings of the 1st international workshop on distributed event-based systems (DEBS'02).* Vienna, Austria.

328. Peter Triantafillou and Andreas Economides. 2004. Subscription summarization: A new paradigm for efficient publish/subscribe systems. In *ICDCS,* 562–571. Tokyo, Japan.

329. Jonathan S. Turner, Patrick Crowley, John DeHart, Amy Freestone, Brandon Heller, Fred Kuhns, Sailesh Kumar, et al. 2007. Supercharging Planetlab: A high performance, multi-application, overlay network platform. *SIGCOMM Comput. Commun. Rev.* 37(4):85–96.

330. UPnP Forum. 2000. UPnP device architecture. http://www.upnp.org/download/UPnPDA10_20000613.htm,

331. Robbert Van Renesse, Kenneth P. Birman, and Werner Vogels. 2003. Astrolabe: A robust and scalable technology for distributed system monitoring, management, and data mining. *ACM Trans. Comput. Syst.* 21(2):164–206.

332. Robbert van Renesse, Dan Dumitriu, Valient Gough, and Chris Thomas. 2008. Efficient reconciliation and flow control for anti-entropy protocols. In *LADIS '08: Proceedings of the 2nd workshop on large-scale distributed systems and middleware,* 1–7, White Plains, NY: ACM.

333. Antonio Virgillito, Roberto Beraldi, and Roberto Baldoni. 2003. On event routing in content-based publish/subscribe through dynamic networks. In *Proceedings of the 9th IEEE workshop on future trends of distributed computing systems (FTDCS 2003).* 322–328: IEEE.

334. Aggelos Vlavianos, Marios Iliofotou, and Michalis Faloutsos. 2006. BiToS: Enhancing BitTorrent for supporting streaming applications. In *25th IEEE international conference on the IEEE computer communications, joint conference of the IEEE computer and communications societies* (23–29 Apr.). Barcelona, Catalunya.

335. Spyros Voulgaris, Daniela Gavidia, and Maarten van Steen. 2005. CYCLON: Inexpensive membership management for unstructured P2P overlays. *J. Network Syst. Manage.* 13(2).

336. Spyros Voulgaris, Etienne Rivire, Anne-Marie Kermarrec, and Maarten Van Steen. 2006. Sub-2-Sub: Self-organizing content-based publish subscribe for dynamic large scale collaborative networks. In *In IPTPS06: The 5th international workshop on peer-to-peer systems.* Santa Barbara, CA.

337. Litwin W. and T. Risch. 1997. LH*g: A high-availability scalable distributed data structure by record grouping. *IEEE Transactions on Knowledge and Data Engineering,* 14(4): 923–927.

338. Marc Waldman, Aviel D. Rubin, and Lorrie Faith Cranor. 2000. Publius: A robust, tamper-evident, censorship-resistant web publishing system. In *SSYM'00: Proceedings of the 9th conference on USENIX security symposium,* 5–5, Berkeley, CA: USENIX Association.

339. Michael Walfish, Hari Balakrishnan, and Scott Shenker. 2004. Untangling the web from DNS. In *NSDI'04: Proceedings of the 1st conference on symposium on networked systems design and implementation,* 225–238, San Francisco, CA: USENIX Association.

340. Guohui Wang, Bo Zhang, and T. S. Eugene Ng. 2007. Towards network triangle inequality violation aware distributed systems. In *IMC '07: Proceedings of the 7th ACM SIGCOMM conference on internet measurement,* 175–188, New York: ACM.

341. Jiajun Wang, Cheng Huang, and Jin Li. 2008. On ISP-friendly rate allocation for peer-assisted vod. In *MM '08: Proceedings of the 16th ACM international conference on multimedia,* 279–288, New York: ACM.

342. Limin Wang, Vivek Pai, and Larry Peterson. 2002. The effectiveness of request redirection on CDN robustness. In *OSDI '02: Proceedings of the 5th symposium on operating systems design and implementation,* 345–360, New York: ACM.

343. Limin Wang, Kyoung Soo Park, Ruoming Pang, Vivek Pai, and Larry Peterson. 2004. Reliability and security in the CoDeeN content distribution network. In *ATEC '04: Proceedings of the annual conference on USENIX annual technical conference,* 14–14, Berkeley, CA: USENIX Association.

344. Yi-Min Wang, Lili Qiu, Dimitris Achlioptas, Gautam Das, Paul Larson, and Helen J. Wang. 2002. Subscription partitioning and routing in content-based publish/subscribe networks. In D. Malkhi, ed., *Distributed algorithms,* vol. 2508/2002 of *Lecture notes in computer science,* Oct.

345. Yi-Min Wang, Lili Qiu, Chad Verbowski, Dimitris Achlioptas, Gautam Das, and Paul Larson. 2004. Summary-based routing for content-based event distribution networks. *SIGCOMM Comput. Commun. Rev.* 34(5):59–74.

346. Duncan J. Watts. 1999. *Small worlds: the dynamics of networks between order and randomness.* Princeton, NJ: Princeton University Press.

347. A. Whitaker and D. Wetherall. 2002. Forwarding without Loops in Icarus. In *Open Architectures and Network Programming*, 63–75.

348. Pawel Winter. 1987. Steiner problem in networks: A survey. *Netw.* 17(2):129–167.

349. Chuan Wu, Baochun Li, and Shuqiao Zhao. 2007. Magellan: Charting large-scale peer-to-peer live streaming topologies. In *ICDCS '07: Proceedings of the 27th international conference on distributed computing systems*, 62, Washington, DC: IEEE Computer Society. Toronto, Canada.

350. Jun Xu, Abhishek Kumar, and Xingxing Yu. 2003. On the fundamental tradeoffs between routing table size and network diameter in peer-to-peer networks. *IEEE J. Select. Areas Commun.* 22:151–163.

351. Beverly Yang and Hector Garcia-Molina. 2003. Designing a super-peer network. *International conference on data engineering*: 49. Banglore, India

352. Xiangying Yang and Gustavo de Veciana. 2004. Service capacity of peer to peer networks. In *Proceedings of IEEE INFOCOM*, 2242–2252. Hongkong, China.

353. Xiangying Yang and Gustavo de Veciana. 2006. Performance of peer-to-peer networks: Service capacity and role of resource sharing policies. *Perform. Eval.* 63(3):175–194.

354. Xiaowei Yang, David Clark, and Arthur W. Berger. 2007. NIRA: A new inter-domain routing architecture. *IEEE/ACM Trans. Netw.* 15(4):775–788.

355. Haifeng Yu, Michael Kaminsky, Phillip B. Gibbons, and Abraham D. Flaxman. 2008. SybilGuard: Defending against Sybil attacks via social networks. *IEEE/ACM Trans. Netw.* 16(3):576–589.

356. Y. Yue, C. Lin, and Z. Tan. 2006. Analyzing the performance and fairness of BitTorrent-like networks using a general fluid model. *Comp. Commun.* 29(18):3946–3956.

357. Andreas Zeidler and Ludger Fiege. 2003. Mobility support with REBECA. In *ICDCS workshops*. Providence, RI.

358. Hui Zhang, Ashish Goel, and Ramesh Govindan. 2003. Incrementally improving lookup latency in distributed hash table systems. *SIGMETRICS Perform. Eval. Rev.* 31(1):114–125.

359. Hui Zhang, Ashish Goel, and Ramesh Govindan. 2004. Using the small-world model to improve Freenet performance. *Comput. Netw.* 46(4):555–574.

360. Xinyan Zhang, Jiangchuan Liu, Bo Li, and Tak-Shing Peter Yum. 2005. CoolStreaming/DONet: A data-driven overlay network for peer-to-peer live media streaming. In *Proc. IEEE INFOCOM*, Mar., Miami, FL.

361. Zhan Zhang and Shigang Chen. 2006. Capacity-aware multicast algorithms on heterogeneous overlay networks. *IEEE Trans. Parallel Distrib. Syst.* 17(2):135–147.

362. Ben Y. Zhao, Ling Huang, Jeremy Stribling, Sean C. Rhea, Anthony D. Joseph, and John D. Kubiatowicz. 2004. Tapestry: A resilient global-scale overlay for service deployment. *IEEE J. Select. Areas Commun.* 22:41–53.

363. Ben Y. Zhao, John D. Kubiatowicz, and Anthony D. Joseph. 2002. Tapestry: A fault-tolerant wide-area application infrastructure. *SIGCOMM Comput. Commun. Rev.* 32(1):81–81.

364. Dong Zhou, Yuan Chen, Greg Eisenhauer, and Karsten Schwan. 2001. Active brokers and their runtime deployment in the ECho/JECho distributed event systems. In *Active Middleware Services*, 67–72. San Francisco, CA.

365. Shelley Zhuang, Ben Y. Zhao, Anthony D. Joseph, Randy H. Katz, and John D. Kubiatowicz. 2001. Bayeux: An architecture for scalable and fault-tolerant wide-area data dissemination. In *The 11th international workshop on network and operating systems support for digital audio and video (NOSSDAV'01)*, (Jun.), 11–20, Port Jefferson, NY.

366. Shelley Zhuang, Dennis Geels, Ion Stoica, and Randy H. Katz. 2005. On failure detection algorithms in overlay networks. In *INFOCOM*, 2112–2123. Miami.

Index

A

Access control, 186–188
Accountable Internet Protocol (AIP), 22
Acyclic routing topologies, 145
    mobility support, 163
Addressing system, 18–19, *See also* IP addresses
Advertisement inheritance routing, 157
Advertisement semantics, 147, 152, 156
    Hermes rendezvous point system, 157–158
Akamai, 8, 206, 207–208, 211–212
Amazon Dynamo, 85, 130, 189–197, 214
    architecture, 191–192
    consistent hashing, 191, 193–194
    data replication and versioning, 194–195
    failure detection, 197
    partitioning algorithm, 193–194
    quorum-like techniques, 191, 196
    relational databases, 190
    ring structure, 193
    vector clocks, 195–196
    virtual nodes, 194
    Web services stack, 189–190
Anonymous content storage, 8
Anonymous routing and communications, 180
    censorship resistance, 173–174
    censorship resistance, Achord, 184
    Hordes, 184–185
    jondo process, 184
    Mist hierarchical routing system, 185–186
    mixes, 180
    onion routing, 6, 8, 181, 182, 184
    P2P anonymization system (formerly Tarzan),
        182–184
    Tor, 181–182
Anti-entropy protocols, 131, 132–133
Antisnubbing, 55
Anycasting, 5, 14
Application-layer multicast, 24–25
Application programming interface (API), 15
    Coral CDN, 209
    DHT, 86
    Freenet system, 65–66
    Pastry, 93
Applications layer, 2, 4
Approximate concurrent state machine
        (ACSM), 125
AstroLabe, 130
Attacks and threats, 165–169, 174–175

denial of service, 56, 168, 176
    insider attacks, 176
    outsider attacks, 176–177
Attenuated Bloom filter, 128
Auditing gateway, 145
Authentication, 174
    pub/sub systems, 186–188
Autonomous system (AS), 14
    categories, 21
    self-certifying identifier, 22

B

Bamboo, 96
Barabási-Albert network growth model, 39
BASS, 198–199
Bayeux, 98, 137, 139
Biased neighbor selection, 59
Bigtable, 128
Binary decision diagrams (BDDs), 148, 155
BiToS, 199
BitTorrent, 2, 8, 43, 50–58, 213
    antisnubbing, 55
    broadcasting, 51
    choking, 54–55
    comparisons, 67
    cross-ISP, 58–60
    DHT applications, 55, 85
    end game, 55
    firewalls and, 16
    flash crowd resistance, 51, 56
    HTTP vs., 51
    interdomain routing, 20
    ISPs vs., 56, 58
    Kademlia and Mainline DHT, 55, 101, *See
        also* Kademlia
    key interactions, 52
    networking, 54
    performance evaluation, 57–58
    piece selection, 53
    tit for tat peer selection, 53
    torrent file, 51
    trackerless operation, 55
    vulnerabilities, 56
BitTorrent-Assisted Streaming System
        (BASS), 198–199
Blizzard, 2

235

Printed and bound by CPI Group (UK) Ltd, Croydon, CR0 4YY

23/10/2024

01777686-0006